T0135559

Mechanisms underlying the robustness of oscillatory properties

D I S S E R T A T I O N

zur Erlangung des akademischen Grades
doctor rerum naturalium
(Dr. rer. nat.)

im Fach Biophysik

eingereicht an der

Lebenswissenschaftlichen Fakultät
der Humboldt-Universität zu Berlin

von

Dipl.-Math. Katharina Baum
geboren am 15.04.1984 in Berlin

Präsident der Humboldt-Universität zu Berlin: Prof. Dr. Jan-Hendrik Olbertz
Dekan der Lebenswissenschaftlichen Fakultät: Prof. Dr. Richard Lucius

Gutachter/innen:

1. Prof. Dr. Dr. h.c. Edda Klipp
2. Dr. Jana Wolf
3. Prof. Dr. Nils Blüthgen

Tag der mündlichen Prüfung: 10.09.2014

Bibliografische Information der Deutschen Nationalbibliothek

Die Deutsche Nationalbibliothek verzeichnet diese Publikation in der
Deutschen Nationalbibliografie; detaillierte bibliografische Daten sind
im Internet über http://dnb.d-nb.de abrufbar.

ISBN 978-3-8325-3831-6

Logos Verlag Berlin GmbH
Comeniushof, Gubener Str. 47,
10243 Berlin
Tel.: +49 (0)30 42 85 10 90
Fax: +49 (0)30 42 85 10 92
INTERNET: http://www.logos-verlag.de

Zusammenfassung

Schwingungen, auch Oszillationen genannt, treten im Zusammenhang mit vielen biologischen Prozessen auf: in genetischen Systemen, in der Signaltransduktion und als metabolische Oszillationen. Beispiele sind zirkadiane Oszillationen, die kanonische NF-κB-Signalübertragung und intrazelluläre Kalziumoszillationen. Die Oszillationen unterscheiden sich nicht nur in ihrer Einbettung in unterschiedliche biologische Kontexte und in ihren Zeitskalen, sondern auch in der Stärke ihrer Reaktion auf Störungen, der sogenannten Robustheit oder Sensitivität. Die Periode von Kalziumoszillationen, deren Funktion vermutlich in der frequenzkodierten Signalübertragung besteht, ist dafür bekannt, besonders sensitiv zu sein. Die Periode von zirkadianen Oszillationen ist hingegen besonders robust; sie muss nahezu unbeeinflusst von Temperaturänderungen, Veränderungen des pH-Wertes und des Nährstoffangebots bleiben, um eine zuverlässige Zeitmessung sicherzustellen. A priori sind die Ursachen dieser Unterschiede in der Robustheit der Periode größtenteils unbekannt.

In dieser Doktorarbeit werden Methoden der mathematischen Modellierung dazu verwendet, potenzielle strukturelle Ursachen von Sensitivitätsunterschieden der oszillatorischen Eigenschaften Periode und Amplitude zu untersuchen. Die biologischen oszillierenden Prozesse werden durch Modelle gewöhnlicher Differentialgleichungen repräsentiert und die Analyse beschränkt sich auf Grenzzyklusoszillationen. Die Perioden- und Amplitudensensitivitäten werden mittels eines Maßes der metabolischen Kontrollanalyse bestimmt, wobei endliche statt infinitesimaler Störungen der kinetischen Parameter betrachtet werden. Da die genauen Parameterwerte für die Modelle unbekannt sind, ist die Sensitivitätsanalyse jedes Modells an eine Methode der zufälligen Auswahl der Parameterwerte gekoppelt. Durch diese Vorgehensweise ist es möglich zu unterscheiden, ob Effekte auf die untersuchten oszillatorischen Eigenschaften durch die zugrunde liegende Modellstruktur hervorgerufen werden oder durch die Auswahl der speziellen Parameterwerte. Mithilfe dieser Analysemethode werden zunächst die Sensitivitäten jeweils eines exemplarischen Modells für zirkadiane und für Kalziumoszillationen untersucht. Ihre Perioden- und Amplitudensensitivitäten zeigen deutliche Unterschiede, was die Wahl der Analyse bekräftigt: Das Modell der zirkadianen Oszillationen weist sehr geringe Periodensensitivitäten auf, während die Sensitivitäten der Perioden des Modells der Kalziumoszillationen höher und variabler sind.

In verschiedenen Publikationen wurde bereits der Effekt von Rückkopplungsmechanismen auf die Perioden- und Amplitudensensitivitäten untersucht. Die vorliegende Doktorarbeit erweitert diese Untersuchungen dahingehend, dass nicht nur der Effekt der Art der Rückkopplungen auf die Sensitivität von Periode und Amplitude von oszillierenden Systemen analysiert wird, sondern zusätzlich die Effekte von Eigenschaften des Masseflusses und der Reaktionskinetiken. Zu diesem Zweck wird ein Kettenmodell der Länge vier als Basismodell für Oszillationen verwendet, welches sowohl für eine einzelne negative Rückkopplung als auch für eine einzelne positive Rückkopplung ungedämpfte Grenzzyklusoszillationen aufweist.

In Übereinstimmung mit veröffentlichten Ergebnissen anderer Forschungsgruppen, die andere Methoden und Sensitivitätmaße verwenden, zeigt die hier durchgeführte Analyse des Kettenmodells mit negativer Rückkopplung niedrige Periodensensitivitäten und variable Amplitudensensitivitäten, während das Modell mit positiver Rückkopplung höhere, variable Periodensensitivitäten und geringere, weniger variable Amplitudensensitivitäten aufweist. Unterbrechungen des Masseflusses zwischen zwei aufeinanderfolgenden Spezies im Kettenmodell können die Perioden- und Amplitudensensitivitäten durch Umverteilung der Sensitivitäten auf weniger Parameter oder Verlust kompensatorischer Regulationen erhöhen. Bei Verwendung von sättigenden Kinetiken, die phänomenologisch mittels Michaelis-Menten-Kinetiken modelliert werden, werden im Vergleich zu nichtsättigenden Kinetiken erhöhende Effekte auf die Perioden- und Amplitudensensitivitäten beobachtet. Dies entspricht Ergebnissen dieser Arbeit für Konzentrationen von Spezies im Fließgleichgewicht, nach welchen diese auch schon für nur eine sättigende Reaktion ultrasensitiv gegenüber Parameteränderungen sein können. In den Kettenmodellen kann der Sättigungsgrad verschiedener Reaktionen unterschiedliche Auswirkungen auf die Sensitivitäten von Periode und Amplitude haben.

Die Bedeutung der betrachteten Kinetiken für die Perioden- und Amplitudensensitivitäten wird mit der Untersuchung eines anderen Basis-Oszillators unterstrichen. In einer früheren Publikation wurde bereits mit der Hilfe dieses Oszillators der Einfluss von der Art der Rückkopplung und seiner Stärke untersucht. In der vorliegenden Dissertation wird für dieses Modell festgestellt, dass sättigende Kinetiken in Abbaureaktionen die Perioden- und Amplitudensensitivitäten in gleicher Weise ändern wie die Einführung einer positiven autoregulatorischen Rückkopplung.

Zusätzlich wird in dieser Arbeit die Gültigkeit der mittels der Kettenmodelle gewonnenen Erkenntnisse für weitere, veröffentlichte Modelle von Kalzium- und zirkadianen Oszillationen untersucht. Für einige Modelle ist es dabei möglich, von ihren Perioden- und Amplitudensensitivitäten auf strukturelle Eigenschaften zu schließen. Je stärker ein Modell jedoch von der Struktur eines Kettenmodells abweicht, d.h. je mehr Rückkopplungen und andere strukturelle Motive wie zum Beispiel Erhaltungsbedingungen vorliegen, desto weniger Schlussfolgerungen können auf der Basis der Resultate für die Kettenmodelle gezogen werden. Weiterhin wird der Effekt von sättigenden Kinetiken auf synthetische Oszillatoren als potentielle Anwendung der Ergebnisse dieser Doktorarbeit angeführt. Gleichermaßen könnte die in dieser Arbeit vorgeschlagene Sensitivitätsanalyse zusammen mit der Zufallsauswahl der Parameterwerte genutzt werden, um zu bestimmen unter welchen Bedingungen andere oszillatorische Prozesse, wie zum Beispiel die kanonische NF-κB-Signalübertragung, niedrige oder hohe Perioden- oder Amplitudensensitivitäten aufweisen. Schließlich werden potenzielle Konflikte zwischen den Robustheiten verschiedener Oszillationseigenschaften und mögliche Erweiterungen des Ansatzes diskutiert.

Insgesamt liefert diese Doktorarbeit eine umfassende Analyse des Einflusses verschiedener struktureller Eigenschaften auf die Perioden- und Amplitudensensitivitäten von oszillierenden Prozessen. Sie zeigt, dass nicht nur Rückkopplungen eine wichtige Rolle spielen, sondern dass weitere Eigenschaften des Masseflusses sowie im Besonderen auch die Wahl der Kinetiken von zentraler Bedeutung sind.

Summary

Oscillations occur in many different biological processes: in genetic systems, in signaling and as metabolic oscillations. Examples are circadian oscillations, the canonical NF-κB-pathway, and calcium signaling. The oscillations differ not only in their biological backgrounds and time-scale, i.e. period, but also in the intensity of their response towards perturbations, the so-called robustness or sensitivity of the particular response. Thereby, the period of calcium oscillations whose function is discussed to lie in frequency encoded signal transduction is known to be very sensitive. Contrariwise, the period of circadian rhythms is very robust; it has to remain nearly unaffected by changes of temperature, pH and nutritional conditions in order to provide reliable timing. *A priori*, the origins of these differences in period sensitivities are widely unknown.

In this thesis, mathematical modeling is used to investigate potential structural sources of differences in the sensitivities of the oscillatory properties period and amplitude. As mathematical representation of the biological oscillatory processes, ordinary differential equation models are employed, and it is focused on limit cycle oscillations. The period and amplitude sensitivities are estimated using a measure from metabolic control analysis applying finite perturbations of the kinetic parameters. Since the exact values of the parameters of the models are unknown, the sensitivity analysis of each model is combined with a random sampling procedure in order to distinguish effects imposed by the structure from effects imposed by the choice of the particular parameter set. Using this method, first, the sensitivities of an exemplary circadian rhythm model and of an exemplary calcium oscillations model are analyzed. Their period and amplitude sensitivities are found to be distinct from each other which supports the chosen approach: The circadian rhythm model exhibits very low period sensitivities while the calcium oscillations model higher, variable period sensitivities.

In previous publications, the effect of the feedback type on period and amplitude sensitivities has been examined. This thesis extends those investigations by considering not only the effects of feedbacks, but also those of matter flow properties and kinetics on the period sensitivity and amplitude sensitivity of oscillatory systems. For this purpose, a chain model of length four is employed as basic oscillator model which displays sustained oscillations yielding a negative feedback only as well as exhibiting a positive feedback only.

In accordance with published results using different methods and sensitivity measures, the model with negative feedback displays low period sensitivities and variable amplitude sensitivities, while the positive feedback model yields higher, variable period sensitivities and lower, less variable amplitude sensitivities. Lacking matter flow between species can increase the period and amplitude sensitivities via redistribution of the sensitivities among fewer parameters or lack of compensatory effects. Saturating kinetics which are modeled purely phenomenologically by Michaelis-Menten kinetics reveal to increase the period and amplitude sensitivities. This is in accordance with the finding of a supplemental criterion for ultrasensitivity of the steady state requiring only saturation in one reaction in the presented work. In the chain models, the saturation levels of different reactions can exhibit different impacts on the sensitivities.

The importance of the type of kinetics considered for the period and amplitude sensitivities are re-emphasized using a different basic oscillator model. With its help, the effect of the feedback type and its strength has been investigated in literature. In the presented work, it is found that in this model, saturating kinetics in degradation reactions alter the period and amplitude sensitivities similarly to the addition of an autoregulatory positive feedback loop.

Furthermore, the validity of the results obtained for the chain models is examined for further published models of calcium and circadian oscillations. For some of the models, it is possible to infer their structural properties from the results of a sensitivity analysis with the knowledge derived from the chain models. However, the less a model resembles a chain, i.e. the more feedback loops and other design principles as for example conservation conditions are included, the less conclusions can be drawn on the basis of the results from the chain models. Additionally, the effect of kinetics on synthetic oscillators is proposed as potential application of the results of this thesis. The sensitivity analysis combined with the random sampling approach could also be used to determine under which conditions other oscillatory processes, for example canonical NF-κB signal transduction, exhibit low or high period or amplitude sensitivities. Finally, subjects such as potential robustness trade-offs and further extensions of the approach are discussed.

Taken together, this thesis provides a comprehensive analysis of the impact of different structural features on the period and amplitude sensitivities of oscillatory processes and reveals that not only the feedback type is important, but that also matter flow properties and, in particular, the choice of the kinetics play an important role.

Symbols and abbreviations

90%R	middle 90% data range
A	mean amplitude
J	Jacobian matrix of an ODE system
K_i	nl-parameter being a Michaelis-Menten constant
k_i	rate coefficient
kn_i	nl-parameter
μ	median
N	stoichiometric matrix
n_i	cooperativity parameter
Q_p	pth percentile
R_A^p	amplitude sensitivity coefficient for parameter p
R_T^p	period sensitivity coefficient for parameter p
R_S	Spearman's rank correlation coefficient
$S(0)$	initial species concentration
S^0	steady state species concentration
σ_A	amplitude sensitivity measure
σ_T	period sensitivity measure
T	period
ν	flow vector
V_i	rate coefficient if being a maximal reaction velocity

ATP	adenosine triphosphate
cAMP	cyclic adenosine monophosphate
CI	confidence interval
DNA	desoxyribonucleic acid
ER	endoplasmic reticulum
HB	Hopf bifurcation
IP_3	inositol 1,4,5-triphosphate
LC	limit cycle
MAPK	mitogen activated protein kinase
MCA	metabolic control analysis
mRNA	messenger ribonucleic acid
MWU	Mann-Whitney-U
NF-κB	nuclear factor κ-light-chain-enhancer of activated B-cells
ODE	ordinary differential equation
SCN	suprachiasmatic nucleus

Contents

Zusammenfassung   **i**

Summary   **v**

Symbols and abbreviations   **ix**

Contents   **xi**

1. Introduction   **1**
   1.1. Biological oscillations 1
   1.2. Robustness and sensitivity 2
   1.3. The objectives and the approach 3
      1.3.1. Robustness of the steady state 3
      1.3.2. Robustness of oscillations 4
      1.3.3. The approach used in this work 5
      1.3.4. The approach in the context of published results 7

2. Robustness measure and methods   **9**
   2.1. Mathematical models of biological oscillations 9
      2.1.1. ODE systems 9
      2.1.2. Rate coefficients, nl-parameters, cooperativity parameters . . . 10
      2.1.3. Visualization of ODE models as networks 11
      2.1.4. Determining feedback loops by the Jacobian matrix 14
      2.1.5. Steady states and stability 17
      2.1.6. Limit cycle oscillations 18
      2.1.7. Bifurcation diagrams and Hopf bifurcations 20
   2.2. Quantifying sensitivity: The chosen measures 22

2.3. Sensitivity estimation 26

  2.3.1. Parameter sampling 28

  2.3.2. Determining the stability of the steady state and performing
  model simulations 30

  2.3.3. Checking for regular oscillations and determining period and
  amplitude 31

  2.3.4. Calculation of the sensitivity 32

  2.3.5. Oscillation probability 32

2.4. Methods of statistical analysis 33

  2.4.1. Median, quartiles, 90% data range 33

  2.4.2. Confidence interval of the median 35

  2.4.3. Box-plot visualization 36

  2.4.4. Spearman's rank correlation coefficient R_S 37

  2.4.5. Testing the significance of differences 38

2.5. Summary and discussion: Robustness measure and methods 42

  2.5.1. ODE models and oscillation detection 42

  2.5.2. Mean amplitude as amplitude characteristic 43

  2.5.3. Sensitivity measure 44

  2.5.4. Sampling approach and sensitivity estimation 45

  2.5.5. Statistical evaluation and data interpretation 47

3. The influence of the structure on robustness   49

3.1. Circadian and calcium oscillations 49

  3.1.1. Circadian oscillations 49

  3.1.2. Calcium oscillations 51

  3.1.3. Differences between circadian and calcium oscillations 53

3.2. Model structure or choice of the parameter set 54

  3.2.1. The chosen models 56

  3.2.2. Results of the sensitivity analysis 56

3.3. Summary and discussion: The influence of the structure on robustness 60

  3.3.1. Structural properties to be examined in this work 62

4. The effect of feedbacks on sensitivity   **63**

4.1. Feedbacks and feedback loops in general 63

    4.1.1. Feedbacks and feedback loops in biology 63

    4.1.2. Known effects of feedbacks on oscillatory properties 64

    4.1.3. Feedback vs. feedback loop in ODE systems 67

4.2. The chain models 68

    4.2.1. Implementation of the feedback 70

    4.2.2. Stability analysis and fixation of the Hill coefficients 72

4.3. Sensitivity analysis for different feedback types 74

4.4. Further investigation of the sensitivities of the chain models 77

    4.4.1. Branching in the period sensitivity of the negative feedback
           chain model 77

    4.4.2. Bifurcation analyses of the negative feedback chain model . . 79

    4.4.3. Bifurcation analyses and birhythmicity in the positive feed-
           back chain model 82

4.5. Summary and discussion: The effect of feedbacks on sensitivity 85

    4.5.1. Relation to other findings 86

    4.5.2. Influence of the model assumptions on sensitivity 89

    4.5.3. Oscillation probability 91

    4.5.4. More possibilities of feedback implementation 92

5. The effect of matter flow on sensitivity   **95**

5.1. Matter flow: conversion vs. regulated production 95

    5.1.1. Matter flow vs. flow of information in biology 98

5.2. Matter flow in the chain models 100

    5.2.1. Matter flow properties influence the regulatory structure . . . 101

    5.2.2. Is a further simplification of the analysis possible? 102

5.3. Sensitivity analysis for different matter flow properties 103

5.4. Origins of particular effects of matter flow on sensitivity 105

    5.4.1. Increase of sensitivities in the negative feedback chain models 105

    5.4.2. Decrease in dispersion of the period sensitivity in the negative
           feedback chain models 108

    5.4.3. Increase of sensitivities in the positive feedback chain models . 110

Contents

5.5. Summary and discussion: The effect of matter flow on sensitivity . . 113

    5.5.1. Matter flow from more than one source species 115

    5.5.2. Lacking matter flow in signaling processes 116

    5.5.3. Applicability of the results 116

6. The effect of saturating kinetics on sensitivity   **117**

  6.1. Mass action kinetics vs. Michaelis-Menten kinetics and saturation . . 118

    6.1.1. Mass action kinetics 118

    6.1.2. Michaelis-Menten kinetics 118

    6.1.3. Saturation 119

    6.1.4. The meaning of Michaelis-Menten kinetics in this work 122

    6.1.5. Known effects of saturating kinetics on oscillations 123

  6.2. The effect of saturating kinetics on steady state sensitivity 124

    6.2.1. Zero-order ultrasensitivity in the Goldbeter-Koshland switch . 125

    6.2.2. Sensitivity of the steady state and ultrasensitivity: The response coefficient R_ν 127

    6.2.3. Re-examination of the Goldbeter-Koshland switch 129

    6.2.4. Open switch: The Goldbeter-Koshland switch with production and degradation 131

    6.2.5. Short chain: The open switch without backward reaction . . . 135

    6.2.6. Enhanced sensitivity of the steady state 138

  6.3. The effect of saturating kinetics on period and amplitude sensitivity . 138

    6.3.1. Michaelis-Menten kinetics in the chain model 139

    6.3.2. Sensitivity analysis for different kinetics 140

    6.3.3. Influence of the K_M-values 143

    6.3.4. The influence of the number of saturated reactions 145

    6.3.5. Influence of particular reactions being saturated 148

  6.4. Saturation and oscillation probability 151

  6.5. Summary and discussion: The effect of saturating kinetics on sensitivity 154

    6.5.1. Saturating kinetics in conversions vs. saturating kinetics in regulated productions 155

    6.5.2. Possible extensions of the examination of kinetics 156

7. Further applications   **159**

7.1. It is not only feedback that matters 159

7.2. Inferring structural properties from sensitivities 163

     7.2.1. Results of the sensitivity analyses 164

     7.2.2. Circadian rhythm models 166

     7.2.3. Calcium oscillations models 171

7.3. Summary, discussion and outlook: Further applications 177

     7.3.1. Applicability of structure prediction 179

     7.3.2. Suggesting structure to tune sensitivity 180

     7.3.3. Determining sensitivity to obtain indications on the biological
          potential of oscillations 181

8. Discussion: Mechanisms underlying the robustness of oscillatory properties   **183**

A. Appendix   **189**

A.1. Robustness measure and methods 189

     A.1.1. Influence of the parameter perturbation on the sensitivity . . . 189

     A.1.2. Influence of the definition of the sensitivity measure on the
          sensitivity results 192

     A.1.3. Influence of the sampling interval on the sensitivity results . . 193

     A.1.4. Values obtained for special rate equations by the sampling
          method 195

     A.1.5. Median vs. arithmetic mean 196

A.2. The effect of feedbacks on sensitivity 198

     A.2.1. Definition of the feedback strength and stability of the steady
          state 198

     A.2.2. Positive feedback term and Hill coefficient 199

     A.2.3. Length of the feedback loop 201

A.3. The effect of matter flow on sensitivity 203

     A.3.1. Transformation of models with different matter flow properties 203

     A.3.2. Equivalent models may have different sensitivities 208

A.3.3. Sensitivity analysis of the negative feedback chain models with
        different matter flow properties 211
A.3.4. Sensitivity analysis of the positive feedback chain models with
        different matter flow properties 217
A.4. The effect of saturating kinetics on sensitivity 220
    A.4.1. Derivation of Michaelis-Menten kinetics 220
    A.4.2. Steady state response coefficients as in the main text 221
    A.4.3. Steady state response coefficients according to MCA 228
    A.4.4. Stability of the chain models with saturating kinetics 234
    A.4.5. Hill coefficients in the chain models with saturating kinetics . 235
    A.4.6. Influence of the definition of a saturated reaction 236
    A.4.7. Particular reactions being saturated 237
    A.4.8. Chain models only partly employing saturating kinetics 240
    A.4.9. Matter flow and saturating kinetics 241
A.5. Further applications 243
    A.5.1. Sensitivity of synthetic oscillators 243
    A.5.2. Sensitivity of an NF-κB model 244
A.6. Tables for sensitivity values 245
    A.6.1. Chain models with positive or negative feedback 245
    A.6.2. Chain models with different matter flow properties 246
    A.6.3. Chain models with saturating kinetics 248
    A.6.4. Models of Tsai et al. and alterations 248
    A.6.5. Circadian and calcium oscillations models 249
    A.6.6. Synthetic oscillators 251
    A.6.7. NF-κB oscillations model 251
A.7. Model equations 252
    A.7.1. Circadian and calcium oscillations models 252
    A.7.2. Chain models of length four with mass action kinetics 261
    A.7.3. Chain models with different matter flow properties 262
    A.7.4. Chain models with saturating kinetics 262
    A.7.5. Chain models of length five 263
    A.7.6. Models according to Tsai et al. [2008] and alterations 263
    A.7.7. Synthetic oscillators 265

A.7.8. NF-κB oscillations model 269

Acknowledgements **285**

1. Introduction

The definitions of systems biology are various, but the goal of systems biology, as stated by Kitano [2007], is to investigate fundamental and structural principles underlying biological processes. This is also the focus of this thesis as it belongs to the field of systems biology. In its center of interest lie biological processes yielding oscillatory dynamics, and in particular their robustness. More specifically, the aim of this thesis is to examine which system properties lead to robust or sensitive oscillatory characteristics as period and amplitude. For this purpose, it employs a computational approach. The oscillatory processes are represented by models of ordinary differential equations (ODEs) on which a sensitivity analysis including a Monte-Carlo random-sampling method is performed. With the help of a prototype oscillator model, effects of different feedback types, matter flow and kinetics on the robustness of period and amplitude are examined.

In this chapter, biological oscillations are introduced (section 1.1). How the notion of robustness is used in general is described in section 1.2. How the robustness of oscillations is examined here and how this issue has been approached by others is introduced in section 1.3.

1.1. Biological oscillations

Oscillations, which are characterized by a timely repetitive behavior, occur in many cellular processes for different types of networks. Classical examples of biological rhythms are circadian oscillations (their mathematical descriptions are reviewed in Leloup [2009], Bordyugov et al. [2013]) which occur on the level of genetic networks and cell-cycle oscillations manifested in oscillations of cyclins (Evans et al. [1983], Goldbeter [1991]). In metabolism, glycolytic oscillations in yeast and muscle cells

(Higgins [1964], Ghosh and Chance [1964]) and peroxidase oscillations (Yamazaki et al. [1965], Nakamura et al. [1969]) are known to occur. Signaling systems are prone to display oscillating behavior, and oscillatory properties as period and amplitude are discussed to be able to carry information (reviewed in Purvis and Lahav [2013]). Examples are calcium oscillations (reviewed in Berridge and Galione [1988], Dupont et al. [2011]), nuclear factor κ-light-chain-enhancer of activated B-cells (NF-κB) oscillations on single cell-level (Nelson et al. [2004], Sung et al. [2009], Ashall et al. [2009]) and oscillations in the mitogen activated protein kinase (MAPK) cascade on single-cell level (Shankaran et al. [2009]). Also neuronal oscillations, i.e. electric potential changes for excitation transfer (measured in Hodgkin et al. [1952]), and oscillations in cyclic adenosine monophosphate (cAMP) concentrations (Geller and Brenner [1978]) are classical examples. Oscillations of the tumour protein 53 - mouse double minute 2 homolog (p53-Mdm2) response to DNA damage have been detected (Lev Bar-Or et al. [2000]). In the last decade, even synthetic oscillators have been designed where systems are artificially wired in order to produce oscillations (Elowitz and Leibler [2000], Fung et al. [2005], Stricker et al. [2008], Tigges et al. [2009], Toettcher et al. [2010], analyzed and compared in Purcell et al. [2010]).

The present work is particularly interested in self-sustained, undamped limit cycle oscillations (for details on the definition of limit cycle oscillations, see section 2.1.6). They are examined using mathematical models.

1.2. Robustness and sensitivity

Robustness is a common term in biology and particularly in systems biology. In the last 17 years, various definitions and explanations of robustness have been published. In a pioneering work, Barkai and Leibler [1997] state "that the key properties of biochemical networks should be robust in order to ensure their proper functioning" meaning that "they are relatively insensitive to the precise values of biochemical parameters". Ma and Iglesias [2002] express the concept of robustness more generally as follows: "By saying that a system is robust we imply that a particular function or characteristic of the system is preserved despite changes in the operating environment." Csete and Doyle [2002] propose that robustness is characterized by "the preservation of particular characteristics despite uncertainty in components or the environment". Stelling et al. [2004a] define robustness to be the "ability to maintain performance in the face of perturbations and uncertainty" or the "persistence of a system's characteristic behavior under perturbations or conditions of uncertainty". Similarly, Novak and Tyson [2008] rephrase robustness to be "the notion that a control system should function reliably in the face of expected perturbations from outside the control system and from inevitable internal fluctuations". In this work, a system characteristic is considered being robust if it is maintained when the system faces environmental changes.

Sensitivity is the contrary of robustness. Being sensitive means that a system displays a certain variability in its characteristic if exposed to environmental perturbations, i.e. that it lacks robustness. Both terms, robustness and sensitivity, are used here.

First studies that systematically unravel robustness properties have examined the bacterial chemotaxis, i.e. the ability to sense concentration gradients of an attractor (Barkai and Leibler [1997]). Therein, it has been shown that adaptation is a robust feature of bacterial chemotaxis for changes in reaction rate constants and enzymatic concentrations. On the contrary, system activity heavily changes already with changing attractor concentration. Hence, the uncommented statement that the bacterial chemotaxis is robust is not true in general, but only certain system features are robust whereas others are not. When working with robustness, one has to be precise about the system and the type of robustness to be examined.

1.3. The objectives and the approach

1.3.1. Robustness of the steady state

The robustness or sensitivity of steady state properties has been subject to intense research. For example, a theorem for the requirements in mass action networks to display absolute concentration robustness in a species, i.e. the concentration of the species is identical in every positive steady state the system might admit for a certain parameter set, has been derived (Shinar and Feinberg [2010]). A structural requirement for the concentration robustness upon parameter perturbation has been proposed (Steuer et al. [2011]). Robustness of biochemical adaptation, which is connected to robustness of the steady state concentration, has been examined experimentally (Barkai and Leibler [1997]), and topological characteristics for robust biochemical adaptation have been systematically explored theoretically (Ma et al. [2009]). Furthermore, a plethora of structural motifs has been found which can result in ultrasensitivity of the steady state response, i.e. a more than linear increase in the steady state concentration upon change in a particular stimulus (reviewed in Zhang et al. [2013]). Examples for such ultrasensitive response motifs are positive feedback (Sneppen et al. [2008]) which is known to produce also bistability (Straube [2013]), multisite phosphorylation (Markevich et al. [2004], Salazar and Höfer [2009]), and zero-order ultrasensitivity in conversion cycles (Goldbeter and Koshland [1981]). The stability of steady states as indicator for steady state robustness has also been subject to numerous studies (Kitano [2007]).

1.3.2. Robustness of oscillations

In contrast to the robustness of steady states, the robustness of transient processes as oscillations and their properties has received less attention. In previous studies, robustness in the context of oscillations most frequently refers to the size and shape of the region in which the system shows oscillations (Kurosawa and Iwasa [2002], Ma and Iglesias [2002], Stelling et al. [2004b], Buchler et al. [2005], Wagner [2005], Wong et al. [2007], Tian et al. [2009], Apri et al. [2010], Xu and Qu [2012]). Also, robustness in the sense of how well an oscillatory model can fit certain oscillatory data or reference curves has been employed and effects of structural properties of the models as kinetics and wiring of feedback loops on this particular type of robustness have been determined (Saithong et al. [2010a,b]).

However, depending on the system, also the period or amplitude of the oscillations may be differently robust to changes in the environment, and this may reflect the function of the particular rhythm. Circadian oscillations endue a time-keeping function and it has been shown that their period of roughly 24 hours is temperature compensated and also does not change with varying pH or nutritional conditions (Pittendrigh [1954], Pittendrigh and Caldarola [1973], Ruby et al. [1999], Ruoff et al. [2000]). Contrarily, the period of calcium oscillations varies from seconds to minutes and is responsive to changes in temperature, and agonist concentration, the latter a phenomenon often referred to as frequency encoding of the stimulus (Rooney et al. [1989], De Koninck and Schulman [1998], Dolmetsch et al. [1998], Cai et al. [2008], Sanderson et al. [2008], Schipke et al. [2008]). Hence, there are two oscillatory biological processes, the one seems to exhibit a very robust period, the other certainly not. The two systems differ among other things in their regulation, time scales, and flow of matter and it is difficult *a priori* to single out which are the properties making the period of one oscillator more robust than the other. Thus, it is of particular interest to examine which structural principles render the period and amplitude robust or sensitive.

1.3.3. The approach used in this work

When investigating robustness it is necessary to determine (i) which system is looked at, (ii) which feature of the system and (iii) which environmental changes are examined (Stelling et al. [2004b]). These three characteristics are specified according to their use in this thesis in the following.

(i) This study focuses on oscillatory processes in biological systems. For this purpose, model approaches such as network analysis, stoichiometric analysis and structural kinetic modeling do not provide enough detail since the oscillatory properties cannot be determined from them (Steuer [2007]). Instead, a mathematical representation by ODE systems is chosen for the oscillatory processes in question (see section 2.1). The aim is to examine limit cycle oscillations, i.e. sustained oscillations which are further described in section 2.1.6. In other model types as for example regulatory network models, advances to determine the effect of coherence of regulation, feed-forward or feedback loops on the robustness of attractors with respect to perturbations of updating rules have been made (Kwon and Cho [2008], Wu et al. [2009], Le and Kwon [2013]).

(ii) The robustness or sensitivity of two features of an oscillatory process are at the center of interest in this work: The period T and the amplitude. Both characteristics of oscillations are visualized in Figure 1.1. The *period T* is defined as the inter-spike interval or, more mathematically, as the value such that $S_i(t + T) = S_i(t)$ for every oscillating species S_i and every time point $t > T_0$, i.e. for the solution

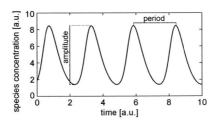

Figure 1.1.: Visualization of oscillatory properties: period and amplitude.

after a certain time the system needs to reach the limit cycle. The *amplitude* is the concentration difference from the largest concentration value and the lowest of one species during one passage through the limit cycle. Since each of the analyzed oscillatory systems consists of more than one species, the arithmetic mean value of the amplitudes (mean amplitude) of all species A is examined.

(iii) The perturbations applied are represented via changes in kinetic parameters as rate coefficients, maximal reaction velocities, inhibition or activation constants and Michaelis-Menten constants (K_M). The characteristics of the mechanistic interactions between species are not varied, meaning that stoichiometry, parameters determining cooperativity as Hill coefficients, and the functional form of the rate equations remain unchanged. These features can be assumed to remain unaffected by perturbations, such as agonist stimulation or temperature variations. In literature, another category of perturbation than parameter perturbation can be found: stochastic noise, which is the random fluctuation of the molecule abundance. It is particularly interesting for small molecule numbers and can be examined with the help of stochastic differential equation models. Since this work focuses on the representation of biological processes by ordinary differential equation systems, the role of stochastic noise for robustness is not explored here. Results for robustness of oscillations towards stochastic noise can be found for example in Gonze et al. [2002, 2008], Gonze and Hafner [2011], Gerard et al. [2012].

To summarize in one sentence: In this work, the robustness of period and amplitude of self-sustainedly oscillating systems against parameter disturbances is examined. It is the aim to investigate which structural characteristics of a biological process render oscillatory properties as period and amplitude robust. Thereby, it is focused on structural characteristics as type of feedback, matter flow and kinetics.

1.3.4. The approach in the context of published results

Sensitivity analyses using a similar measure for the sensitivity as employed here have been performed for particular oscillatory models (e.g. Ihekwaba et al. [2004], Stelling et al. [2004a], Wolf et al. [2005], Bagheri et al. [2007], Wilkins et al. [2007]). Thereby, most frequently circadian oscillations models are chosen (Stelling et al. [2004a], Bagheri et al. [2007], Wilkins et al. [2007]), and/or the main goal is to distinguish between model designs (Stelling et al. [2004a]) or to determine which parameters, or types of parameters, are the most sensitive or robust for the oscillatory process (Ihekwaba et al. [2004], Stelling et al. [2004a], Bagheri et al. [2007], Wilkins et al. [2007]). In contrast, this thesis aims at comparing overall sensitivities of different models. This has been started in Wolf et al. [2005], where the sensitivities of different basic oscillator models at one particular parameter set were compared. This approach does not permit to distinguish whether effects on the sensitivity are caused by the choice of the parameter set or are imposed by the structural model properties. Therefore, in this thesis, the sensitivity analysis is combined with a Monte-Carlo sampling approach. This means that for each model, a large number of parameter sets is randomly sampled and analyzed with respect to period and amplitude robustness (for details of the procedure, see section 2.3.1). The resulting set of period and amplitude sensitivities reveals the robustness or sensitivity potential of each examined model, and enables to compare effects of feedback type, matter flow and kinetics on sensitivity properties.

Other sensitivity analyses have also employed sampling methods, but partly different measures for sensitivity (Dayarian et al. [2009], Hafner et al. [2009], Zamora-Sillero et al. [2011]). Therein, the so-called "viable" region in parameter space of a model has been determined whose size and shape is considered to deliver information on robustness (Dayarian et al. [2009]). In Hafner et al. [2009], the determination of the size of the viable region as "global" measure has been combined with a "local" sensitivity analysis of the parameter sets found in the viable region. Different local robustness quantifiers have been used, including the effect of individual parameter perturbations on the period. Similarly to the approach in this thesis, the authors have compared the obtained distributions of the robustness quantifiers for all parameter sets and have concluded on which model is more robust than another. However, the authors have rather concentrated on determining which model is more valid for a biological process, and have applied their method to two cyanobacterial models; it has not been considered how general structural properties influence the robustness. Zamora-Sillero and colleagues [2011] have examined a general oscillatory model using the viable-region-approach to determine the effects of different feedbacks on the robustness. However, effects caused by each single feedback and by kinetics can hardly be distinguished due to the choice of their examined model.

The effect of feedbacks on the variability of oscillatory properties has been also investigated in other mathematical models (Tsai et al. [2008], Nguyen [2012]) or synthetic oscillators together with the according model (Stricker et al. [2008]). In contrast to quantifying how strongly the period or amplitude is varied upon small changes in each parameter as in this thesis, the authors of these publications have rather determined the "tunability" of an oscillatory property for broad changes in one particularly chosen parameter. However, "tunability" as in the above mentioned publications and "sensitivity" as employed in this work are not necessarily equal, they are rather two facets of the robustness of oscillatory properties. In section 4.1.2, the results obtained in the above mentioned publications are described in detail. Interestingly, it becomes clear that the method of this thesis confirms for the herein employed sensitivity measure the published conclusions on the effects of feedbacks on the tunability (section 4.5.1). In chapter 7, section 7.1, it is shown that the results based on sensitivity or tunability also coincide for the comparison of two of the models proposed by Tsai et al. [2008].

This thesis extends the results from previous publications by investigating not only the effects of feedbacks (chapter 4), but also those of matter flow properties (chapter 5) and kinetics (chapter 6) on the period and amplitude sensitivities of oscillatory systems. Additionally, a criterion for ultrasensitivity of the steady state requiring only saturation in one reaction is found (section 6.2). The applicability of the results for feedbacks and kinetics in the context of sensitivity is investigated in more complex models, and experiments for testing the predictions concerning kinetics and sensitivity are suggested for synthetic oscillators (chapter 7).

2. Robustness measure and methods

The methods used in this work are explained in this chapter. First, notions concerning ordinary differential equation systems which are used to model the biological processes are introduced (section 2.1). Second, the employed measure of period and amplitude robustness is described (section 2.2). Third, the method of sensitivity analysis is step-wise presented (section 2.3). The chapter ends with describing the statistical methods applied on the data generated during the sensitivity analyses (section 2.4) and discussion of the introduced methods (section 2.5).

2.1. Mathematical models of biological oscillations

This study focuses on the theoretical analysis of oscillatory behavior of biological processes. In order to perform a mathematical evaluation, mathematical representations by ordinary differential equation (ODE) systems are chosen for the processes in question. The general theory of ODE systems can be found e.g. in Hubbard and West [1995].

2.1.1. ODE systems

In an ODE system, the m time-dependent species concentrations $S_i(t)$ (also called variables) are in general determined by

$$\frac{\mathrm{d}S_i(t)}{\mathrm{d}t} = \sum_{j=1}^{M} \eta_{ij} \nu_j, \ i = 1, \ldots, m. \tag{2.1}$$

The stoichiometric coefficients η_{ij}, $j = 1, \ldots, M$, $i = 1, \ldots, m$, determine the stoichiometric matrix $N = (\eta_{ij})$. The directions of the reactions are coded in this matrix and the coefficients η_{ij} are assumed to take integer values.

The M enzyme-kinetic rate equations ν_j, $j = 1, \ldots, M$ are also termed *flows*, *reaction rates* or only *reactions*. They are real-valued, positive functions depending on kinetic parameters and the species concentrations. By definition, a concentration can only take non-negative values. If the stoichiometric coefficient η_{ij} has a positive sign, species S_i is referred to as a *product species* of reaction ν_j. If η_{ij} is negative, species S_i is called a *source species* of the reaction. If η_{ij} equals zero, species S_i is not influenced by reaction ν_j. To clarify the notions above, an example ODE system with one species $(m = 1)$ and two reactions $(M = 2)$ is given:

$$\frac{\mathrm{d}S_1(t)}{\mathrm{d}t} = \nu_1 - \nu_2 = k_1 - k_2 \frac{S_1^{n_1}}{S_1^{n_1} + kn_1^{n_1}}. \tag{2.2}$$

The stoichiometric coefficients are $\eta_{11} = 1$ and $\eta_{12} = -1$, the stoichiometric matrix is hence $N = (1 - 1)$. This means that species S_1 is the product species of reaction ν_1 and the source species of reaction ν_2. Reaction ν_1 has no source species. Reactions of this type are called *production reactions* or simply *productions*. Reaction ν_2 has no product species. This type of reactions is called *degradation reaction* or simply *degradation*. The explicit rate equations are given by $\nu_1 = k_1$, $\nu_2 = k_2 \frac{S_1^{n_1}}{S_1^{n_1} + kn_1^{n_1}}$.

2.1.2. Rate coefficients, nl-parameters, cooperativity parameters

In this work, ODE models of different biological processes established by different researchers are examined. For each of the models, a possibly different notation is used in the according publication. Therefore, in order to make the models examined here comparable, this study classifies the kinetic parameters determining the rate equations of an ODE system into three groups: (i) *rate coefficients*, (ii) *nl-parameters*, and (iii) *cooperativity parameters*. In the course of this analysis, the units of the model parameters are neglected as only relative changes are considered.

(i) Each reaction rate has exactly one corresponding *rate coefficient* that enters linearly into the rate equation. In literature, the term rate coefficient is widely used and accepted although notations are inconsistent. This study mainly employs the indexed letter k for this type of parameters, as well as the indexed letter V if the parameter has the function of a maximal reaction velocity for a reaction governed by Michaelis-Menten kinetics. In this thesis, rate coefficients are presumed to take positive real values in order to obtain positive rate equations. In the example system (Equation 2.2), the rate coefficient of ν_1 is k_1, the rate coefficient of ν_2 is k_2.

(ii) Each parameter entering non-linearly into a rate equation without being an exponent is hereafter referred to as *nl-parameter*. In contrast to the established term rate coefficient, the notion of *nl-parameters* does not exist in literature but is defined here. Also parameters entering linearly, but into more than one rate equation are sorted into this group in order to distinguish them from rate coefficients. Michaelis-Menten or equilibrium constants, and volume differences belong to this type of parameters. Reaction rates may depend on none, one or even multiple nl-parameters. In this work, nl-parameters are in general denoted by an indexed kn, but in case of Michaelis-Menten kinetics, the notation of an indexed K may be used for K_M-values; nl-parameters are assumed to take positive real values. In the example ODE system of Equation 2.2, reaction ν_1 lacks an nl-parameter, for reaction ν_2, it is kn_1.

(iii) In this thesis, all parameters occurring as exponents in a rate equation belong to the group of *cooperativity parameters*. An example are Hill coefficients. Cooperativity parameters are not an established category either. In general, the exponents are considered to determine the degree of cooperativity present in a reaction. As for nl-parameters, a reaction rate can depend on none, one or multiple cooperativity parameters. Here, these parameters are denoted by an indexed n and take generally positive integer values if not stated differently. In the example ODE system (Equation 2.2), reaction ν_1 has no cooperativity parameter, for reaction ν_2, the only cooperativity parameter is given by n_1.

2.1.3. Visualization of ODE models as networks

To get an overview on the dependencies in a biological process encoded in an ODE model at a glance, the visualization by nodes and arrows in networks is very helpful. However, it may be misleading due to frequently encountered ambiguities. It can also complicate the communication between experimenters and modelers, but also between modelers using different types of models e.g. ODE models and logical models, since their arrows or even nodes may code for different biological entities. Therefore, a careful description of how the network representations are created and of the meaning of each component is necessary. The visualization procedure employed here is explained with an example ODE system in Figure 2.1 and is described in the following.

In this work, ODE models are visualized by drawing networks consisting of nodes and arrows. Each variable of the ODE system of the model is coded by one node (step 1 in Figure 2.1). Each reaction is coded by exactly one arrow from all its source species to all its product species as defined by the stoichiometric matrix (see section 2.1.1, step 2 in Figure 2.1). This means the number of heads of an arrow is exactly the number of product species of the visualized reaction, the number of tails coincides with the number of source species. Exceptions are production and degradation reactions (e.g. reactions ν_1 and ν_4 in the example system presented in Figure 2.1) which lack a source or a product species, respectively. They are represented by arrows starting in the empty space or by arrows pointing to the empty space, respectively.

Regulations of a species or reaction by another (or itself) occur very often in biological systems, and they are also encoded in the ODE system of a model. In ODE models, regulations from one species to another need to be introduced by species concentrations entering into the rate function of a particular reaction, they cannot directly influence another species as such (in contrast to in logical models). A regulation or dependency is hence visualized by an arrow from a species to an arrow coding for a reaction (step 3 in Figure 2.1). Whether the dependency of a reaction ν_j on a species S_i is positive or negative can be uniquely determined by the derivative of the rate function towards the species, $\frac{\partial \nu_j}{\partial S_i}$. If this derivative yields a value greater than zero, ν_j is regulated positively by S_i. This is marked by an arrow with the normal tip from the node for S_i to the arrow encoding reaction ν_j. If this derivative yields a negative value, ν_j is regulated negatively by S_i. This is marked by an arrow with T-shaped head ($\dashv$) from the node for S_i to the arrow encoding reaction ν_j. If the derivative can take positive or negative values depending on (biologically feasible) species concentrations or parameter values, species S_i has an *ambivalent* effect on reaction ν_j which is encoded by an arrow with circle as head ($\multimap$).

In general, in ODE systems coding a biological process, it is assumed that every reaction depends positively on all of its source species. Hence, positive regulatory arrows from the source species to the arrows representing the according reactions are omitted in this work (step 4 in Figure 2.1). The arrows coding for reactions are considered to also represent positive regulations of the reaction by the source species.

$$\frac{\mathrm{d}S_1}{\mathrm{d}t} = k_1 - k_3 \cdot S_1 \cdot S_2 \cdot S_4$$
$$\frac{\mathrm{d}S_2}{\mathrm{d}t} = k_2 \cdot S_4 - k_3 \cdot S_1 \cdot S_2 \cdot S_4$$
$$\frac{\mathrm{d}S_3}{\mathrm{d}t} = k_3 \cdot S_1 \cdot S_2 \cdot S_4 - k_4 \cdot S_3$$
$$\frac{\mathrm{d}S_4}{\mathrm{d}t} = k_5/(1 + S_3^{n_1}) - k_6 \cdot S_4$$

$$\Leftrightarrow \quad \frac{\mathrm{d}}{\mathrm{d}t} \begin{pmatrix} S_1 \\ S_2 \\ S_3 \\ S_4 \end{pmatrix} = N \cdot \nu = \begin{pmatrix} 1 & 0 & -1 & 0 & 0 & 0 \\ 0 & 1 & -1 & 0 & 0 & 0 \\ 0 & 0 & 1 & -1 & 0 & 0 \\ 0 & 0 & 0 & 0 & 1 & -1 \end{pmatrix} \cdot \begin{pmatrix} k_1 \\ k_2 \cdot S_4 \\ k_3 \cdot S_1 \cdot S_2 \cdot S_4 \\ k_4 \cdot S_3 \\ k_5/(1 + S_3^{n_1}) \\ k_6 \cdot S_4 \end{pmatrix}$$

Step 1: Represent each species by one node.

S_1

S_3 S_4

S_2

Step 2: Represent each reaction ν_j by one arrow from all its source species to all its product species according to N.

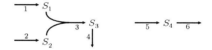

Step 3: Represent each dependency of a reaction by an arrow from the regulating species S_i to the arrow for reaction ν_j:

$$\frac{\mathrm{d}\nu_j}{\mathrm{d}S_i} \begin{cases} = 0 & \text{none} \\ > 0 & \rightarrow \\ < 0 & \dashv \\ \text{changes sign} & \multimap \end{cases}$$

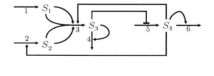

Step 4: Erase arrows representing a dependency of a reaction on one of its source species.

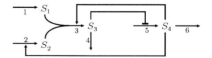

Figure 2.1.: Example for visualizing ODE systems. Perform steps 1 to 3 to obtain an arrow-node visualization of an ODE system as given on top. Step 4 yields a simplification for the visualization of ODE models used in this work which is only applicable to biological systems and is based on the principle that reactions always depend in a positive manner on their source species.

2.1.4. Determining feedback loops by the Jacobian matrix

A feedback loop, also referred to as feedback circuit, is defined as a closed set of oriented interactions (Thomas and Kaufman [2002]).

An interaction between two species exists if the concentration of the first species is affecting the concentration of the second species. The interaction is defined to be directed from the first to the second species. All interactions in the example ODE system from Figure 2.1 (again shown in Figure 2.2 A) are given in Figure 2.2 B. In ODE models, species are coded as variables of the ODE system. The ODE model is completely characterized by its variables, reactions, stoichiometric matrix and choice of the parameters and all interactions between species can be derived from these information. Indeed, it is sufficient to consider the Jacobian matrix $J = (j_{il})_{il}$, $i, l = 1, \ldots, m$ of the ODE model whose entries are defined by

$$ j_{il} = \frac{\partial \frac{dS_i}{dt}}{\partial S_l}. \tag{2.3} $$

J is an $m \times m$-matrix, m denotes hereby the number of variables or species of the ODE system. If considering a particular concentration vector and specific parameter values, an interaction between species S_l and species S_i can unambiguously be detected by looking at entry j_{il} of the Jacobian matrix. If j_{il} is zero, there is no interaction between species S_l and species S_i. If j_{il} is greater than zero, the interaction between S_l and S_i is positive (marked by an arrow with a plus sign in Figure 2.2 B), if j_{il} is smaller than zero, the interaction between S_l and S_i is negative (marked by an arrow with a minus sign in Figure 2.2 B). If one wishes to determine interactions independently from a particular state or from specific parameter values, it might occur that j_{il} can switch signs depending on (biologically feasible) concentration or parameter values. In that case, no unique conclusion is possible on the type of interaction between the species. This may be referred to as *ambivalent* interaction (marked by an arrow with both plus and minus sign in Figure 2.2 B). It can arise from non-monotonous rate functions as well as from two reactions affecting S_i which are both regulated by S_l installing a negative and positive regulation by S_l upon S_i. The Jacobian matrix also reveals whether a species concentration is directly affecting itself. If entry j_{ii} of the Jacobian matrix is non-zero, species S_i

directly influences its own concentration. The sign of the entry is in accordance with the sign of the interaction, of course also ambivalent interactions may exist. Self-influence or self-interaction is always found for species subject to degradation reactions, and in fact also for any source species of a reaction which captures the principle "source species of a reaction positively regulate it" of biological processes. Nevertheless, there can be other examples of direct self-interactions apart from these basic ones, for example in ODE models including representations of autocatalysis, i.e. a species is promoting its own production.

A set of oriented interactions is closed, i.e. it installs a feedback loop, if it forms a circle, for example S_1 interacting with S_2, S_2 interacting with S_3, S_3 interacting with S_1. In a feedback loop, each change in a species concentration indirectly further influences its own concentration, the species feeds back on itself. By definition, feedback loops are shortest loops, which means that in one feedback loop, each species can only occur at most once. Feedback loops may consist only of one species. Feedback loops are disjoint if their sets of participating species are disjoint.

Since being defined by interactions between species, all feedback loops of an ODE model can be determined from its Jacobian matrix. The procedure is explained by Thomas and Kaufman [2002], although therein ambivalent interactions are not considered.

In detail, to find a union of disjoint feedback loops in the ODE system, one has to determine a set of non-zero entries of the Jacobian matrix $\{j_{i_1 l_1}, ..., j_{i_r, l_r}\}$ such that their sets of indices of the first dimension $\{i_1, ..., i_r\}$ is the same as their set of the second dimension $\{l_1, ..., l_r\}$ and both contain each index only once. For example, if $\{j_{31}, j_{43}, j_{14}, j_{22}\}$ is a set of non-zero entries of the Jacobian matrix of an ODE system (as for the one given in Figure 2.2 A), the set of indices for the first dimension is $\{3, 4, 1, 2\}$ which is exactly the same as the set of indices for the second dimension $\{1, 3, 4, 2\}$. However, the underlying feedback loops need to be determined from the thus determined sets of entries by finding minimal such sets. For minimal sets it holds that if taking out any individual element or any subset of elements, the indices for the first dimension and the second dimension are no longer the same. The length of the feedback loop is given by the number of non-zero entries of the Jacobian matrix, the participating species are indicated by their according indices occurring in the set of entries. In the example $\{j_{31}, j_{43}, j_{14}, j_{22}\}$, two minimal sets of non-zero entries can be found: $\{j_{31}, j_{43}, j_{14}\}$ and $\{j_{22}\}$. This indicates that a feedback loop of length three containing species S_1, S_3, S_4 and a feedback loop comprising only species S_2, respectively, exist.

Three types of feedback loops can occur: positive, negative and ambivalent feed-

A

$$\frac{dS_1}{dt} = k_1 - k_3 \cdot S_1 \cdot S_2 \cdot S_4$$
$$\frac{dS_2}{dt} = k_2 \cdot S_4 - k_3 \cdot S_1 \cdot S_2 \cdot S_4$$
$$\frac{dS_3}{dt} = k_3 \cdot S_1 \cdot S_2 \cdot S_4 - k_4 \cdot S_3$$
$$\frac{dS_4}{dt} = k_5/(1 + S_3^{n_1}) - k_6 \cdot S_4$$

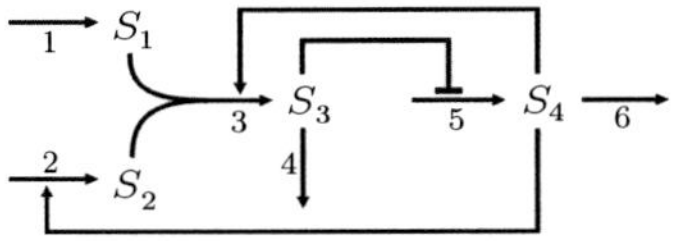

B

$$J = \begin{pmatrix} -k_3 \cdot S_2 \cdot S_4 & -k_3 \cdot S_1 \cdot S_4 & 0 & -k_3 \cdot S_1 \cdot S_2 \\ -k_3 \cdot S_2 \cdot S_4 & -k_3 \cdot S_1 \cdot S_4 & 0 & k_2 - k_3 \cdot S_1 \cdot S_2 \\ k_3 \cdot S_2 \cdot S_4 & k_3 \cdot S_1 \cdot S_4 & -k_4 & k_3 \cdot S_1 \cdot S_2 \\ 0 & 0 & -k_5 \frac{n_1}{(1+S_3^{n_1})^2} & -k_6 \end{pmatrix}$$

o		n		l n
$S_1 \xrightarrow{+} S_3$	$S_1 \xrightarrow{-} S_1$	$S_1 \xrightarrow{-} S_2$	$S_3 \xrightarrow{-} S_4$	$S_4 \xrightarrow{\pm} S_2$
$S_2 \xrightarrow{+} S_3$	$S_2 \xrightarrow{-} S_1$	$S_2 \xrightarrow{-} S_2$	$S_4 \xrightarrow{-} S_4$	
$S_4 \xrightarrow{+} S_3$	$S_4 \xrightarrow{-} S_1$	$S_3 \xrightarrow{-} S_3$		

C

l n

$\{j_{11}\}$, ⊟, $S_1 \xrightarrow{-} S_1$

$\{j_{22}\}$, ⊟, $S_2 \xrightarrow{-} S_2$

$\{j_{33}\}$, ⊟, $S_3 \xrightarrow{-} S_3$

$\{j_{44}\}$, ⊟, $S_4 \xrightarrow{-} S_4$

l n

$\{j_{21}, j_{12}\}$, ⊞, $S_1 \xrightarrow{-} S_2 \xrightarrow{-} S_1$

$\{j_{43}, j_{34}\}$, ⊟, $S_3 \xrightarrow{-} S_4 \xrightarrow{+} S_3$

l n

$\{j_{31}, j_{43}, j_{14}\}$, ⊞, $S_1 \xrightarrow{+} S_3 \xrightarrow{-} S_4 \xrightarrow{-} S_1$

$\{j_{32}, j_{43}, j_{24}\}$, $\mp$, $S_2 \xrightarrow{+} S_3 \xrightarrow{-} S_4 \xrightarrow{\pm} S_2$

l n

$\{j_{31}, j_{43}, j_{24}, j_{12}\}$, $\pm$, $S_1 \xrightarrow{+} S_3 \xrightarrow{-} S_4 \xrightarrow{\pm} S_2 \xrightarrow{-} S_1$

$\{j_{21}, j_{32}, j_{43}, j_{14}\}$, ⊟, $S_1 \xrightarrow{-} S_2 \xrightarrow{+} S_3 \xrightarrow{-} S_4 \xrightarrow{-} S_1$

Figure 2.2.: Example for interactions and feedback loops of an ODE model. A: ODE system of the example model and its visualization. B: Given is the Jacobian matrix, each entry is color-coded with the type of interaction it results in. Green: positive interaction, red: negative interaction, yellow: ambivalent interaction. All interactions between the species are summed up below. C: All feedback loops present in the ODE system from panel A sorted by length. Given are the according set of non-zero entries of the Jacobian matrix, the type of the feedback (positive, negative, ambivalent) and its species- and interaction sequence.

back loops. A feedback loop is positive if the product of all elements from the according set of non-zero entries of the Jacobian matrix is positive. This corresponds to having an even number of negative interactions (and arbitrarily many positive interactions) forming the feedback loop. The feedback loop is negative if the product of all elements from the according set of non-zero entries of the Jacobian matrix is negative which is corresponding to an odd number of negative interactions present in the feedback loop. If the product of all elements from the according set of non-zero entries from the Jacobian matrix can switch sign, i.e. there is one or more ambivalent interaction participating in forming the feedback loop, the feedback loop is referred to as ambivalent.

All feedback loops present in the exemplary system (Figure 2.2 A) are listed together with their type and according sets of non-zero entries of the Jacobian matrix in Figure 2.2 C. A simple numerical routine to determine all feedback loops from the Jacobian matrix of a model with given values for all parameters and species has been implemented in Matlab (Matlab 7.11, The Mathworks, Inc.) for further analysis of the models in chapter 7 (source code is provided on request). Note that for the case of chosen values for species concentrations and parameters, every feedback loop is either positive or negative. The ambivalence of feedback loops can be assessed by varying concentrations or parameter values and checking whether feedback loops can change signs. The connection between the biological understanding of the term "feedback" and feedback loops as defined here is described in section 4.1.3.

2.1.5. Steady states and stability

An important notion in connection with ODE systems is the *steady state*. The steady states of an ODE system are defined to be the solutions (S_i^0) to the possibly non-linear equation system

$$\frac{\mathrm{d}S_i(t)}{\mathrm{d}t} = 0, \; i = 1, \ldots, m. \tag{2.4}$$

Be aware not to confuse the initial species concentration $S_i(0)$ with the notation S_i^0 for the steady state. If the species concentrations are set exactly to a steady state as initial conditions, the solution for the ODE system stays at this steady state. Small deviations from the steady state concentration as initial conditions may suffice to make the solution to the ODE system leave the steady state. If this is the case, the steady state is called unstable. In contrast, a steady state is referred to as stable if for all initial conditions (arbitrarily) close to the steady state, the solution of the ODE system returns back to the steady state. Mathematically, the stability of steady states can be determined with the help of the eigenvalues of the Jacobian matrix $J^0 = (j_{il}^0)_{il}$, $i, l = 1, \ldots, m$, of the ODE system (Equation 2.1) for the steady state concentration vector S^0. In analogy to Equation 2.3, the entries of J^0 are calculated via

$$j_{il}^0 = \frac{\partial \frac{\mathrm{d}S_i}{\mathrm{d}t}}{\partial S_l} |_{S^0}.$$

An eigenvalue to J^0 is a value λ such that there exists a non-zero vector $x \in \mathbb{R}^m$ such that $J^0 \cdot x = \lambda x$. The vector x is then called eigenvector of J^0 to the eigenvalue λ. Eigenvalues can take real or complex values. The steady state is stable if and only if all eigenvalues of J^0 have only negative real parts. Whenever one of the eigenvalues of J^0 exhibits a non-negative real part, the steady state is unstable. Whether a steady state is unstable or not is important for the occurrence of sustained oscillations which can in most cases only be found if also an unstable steady state exists (see also section 2.1.7).

The vector of steady state flows $\nu^0 = \nu(S^0) \in \mathbb{R}^M$ is defined as the vector of the rate equations at steady state species concentration. It can be easily derived that $N \cdot \nu^0 = 0$. This relation is used in the parameter sampling process to obtain the steady state flow vector (section 2.3.1).

2.1.6. Limit cycle oscillations

This study aims to examine the period and amplitude of sustained oscillations (see Figure 1.1). The oscillatory solution of two species of an ODE system with a certain parameter set for explicit initial conditions is illustrated in Figure 2.3 A. The period and amplitude of the oscillation is not the same for each cycle from the beginning of the integration. Instead, only after a certain transient phase (in this case around seven time units), the amplitude and period of the oscillation are approxi-

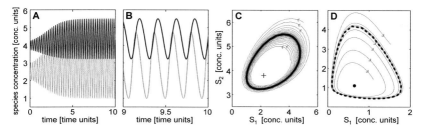

Figure 2.3.: Example for limit cycle oscillations. A: Two species concentrations in time for an oscillatory solution to an ODE system with four variables (S_1: gray, S_2: black). B: Solution from A after the transient phase. C: Phase diagram of two variables of the ODE system from panels A and B. The blue lines give the temporal development of the species concentrations, the arrows indicate the direction of development. The stable limit cycle (LC) is indicated as black line. The unstable steady state is given by the black cross. D: Phase plot of two variables of an ODE system with four variables and unstable LC (black dashed line). The red lines give the temporal development of the species concentrations, the arrows indicate the direction of development. The black dot gives the stable steady state.

mately constant and can be uniquely determined. The behavior after the transient phase is shown in Figure 2.3 B and allows for the determination of the oscillatory characteristics.

In order to be able to determine a unique period and amplitude of an oscillation, it has to satisfy certain conditions which can be visualized in a phase diagram (Figure 2.3 C, D). Here, two substrate concentrations and their development in time, the trajectories, are plotted in phase space, that is the space spanned by the species concentrations. The arrows mark the direction of the temporal development. A region in phase space which attracts all trajectories in its arbitrarily close neighborhood is called an attractor. Cyclic attractors are called stable limit cycles (stable LCs, black circle in Figure 2.3 C). LC oscillations are sustained (i.e. undamped) oscillations with a finite period. Hence, solutions residing on the LC have the property of being oscillations with constant period and amplitude being specific for the parameter values chosen for the underlying ODE system.

Mathematically expressed without considering stability issues, an LC is defined to be a subset of the phase space with the following properties: For any point of a solution to the ODE system residing on the LC $(S_1(t), \ldots, S_m(t)) \in LC$, there exists a finite real value T such that for each species concentration S_i, it is $S_i(t+T) = S_i(t)$ for all time points t. The value T is the period of the LC.

The LC is stable if for initial conditions with any sufficiently small perturbation in either direction, the solution returns to the cyclic attractor. The solution given in Figure 2.3 A, B and C moves toward a stable LC whose portrait in phase space is shown in black in Figure 2.3 C. The blue lines give the trajectories of the ODE system for given initial conditions on which the solution approaches the stable LC. The black cross indicates an unstable steady state.

Unstable LCs occur if the solution leaves the LC for small perturbations as shown in Figure 2.3 D. The solution moves away from the LC given by a dashed black line on the trajectories (red lines). Instead, inside the LC, a stable steady state is marked by a black circle and the solution moves toward it.

For the examination of oscillatory properties in this work, it is focused on stable LC oscillations since they guarantee a unique period and amplitude for given parameter values for the ODE system. However, even for stable LCs, it is difficult to determine the period and amplitude numerically without any calculation error since the temporal development of the species cannot be checked for all points in time (if the work was to be done in finite time) to assure that really the LC is reached. Therefore, this study relies on regular oscillations which are characterized by a sufficiently exact repetitive behavior (for details of the definition of regular oscillations see section 2.3.3).

2.1.7. Bifurcation diagrams and Hopf bifurcations

Bifurcation points are points in parameter space where by small, smooth changes in the parameter values, the qualitative behavior of the solution of the ODE system changes. An example for a change in qualitative behavior is the occurrence of oscillations from a stable steady state. Bifurcations can be detected with the help of bifurcation analyses. In this work, these analyses are performed with the program XPPAUT (Ermentrout [2002]). Typically, only one parameter is varied while the others are kept fixed. The varied parameter is referred to as the bifurcation parameter. The results of bifurcation analyses are visualized in bifurcation

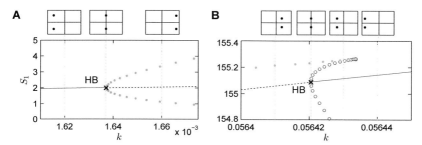

Figure 2.4.: Examples for Hopf bifurcations visualized in bifurcation diagrams of parameter k versus species S_1. A: Supercritical HB. B: Subcritical HB. Coded are: stable steady state (red line), unstable steady state (dashed black line), bifurcation point (HB), maximal and minimal values of species S_1 during the LC oscillations (green dots: stable LC oscillations, blue circles: unstable LC oscillations). The small panels on top show the location of the pair of eigenvalues with switching sign of the real part in the complex plane (x-axis: real part, y-axis: imaginary part, crosses: zeros) for the parameter value indicated by the dashed gray vertical lines.

diagrams which are in this work created with a self-implemented code in Matlab (Matlab 7.11, The MathWorks, Inc). The bifurcation parameter is plotted on the x-axis while one species concentration (or the period) is plotted on the y-axis. For better visualization, different stability regions are marked by different colors and/or symbols as used in XPPAuto and explained in the examples in Figure 2.4.

The most prominent example for a bifurcation through which oscillations arise is the Hopf bifurcation (HB). It is characterized by the transition of the eigenvalues of the Jacobian matrix from a pair of complex eigenvalues with negative real part to a pair of complex eigenvalues with a positive real part. The bifurcation lies at the parameter values where the real part of the examined pair of complex eigenvalues is zero. There are two different types of HBs with slightly different properties. Figure 2.4 A shows the bifurcation diagram for values of the bifurcation parameter k around a *supercritical* HB. The stable steady state (red line) turns unstable (dashed black line). To the right of the bifurcation marked by HB, stable LC oscillations occur which are indicated by green dots. They yield a specific period and amplitude for each specific set of parameter values. The small panels on top show sketches of the values of the pair of complex eigenvalues at values of the bifurcation parameter indicated by the dashed vertical lines. The transition from negative real parts of the eigenvalues (stable steady state) to positive real parts (unstable steady state) passing through the bifurcation can be followed. A *subcritical* HB is given in the bifurcation diagram in Figure 2.4 B. Again, by passing through the HB (from right to left), the stable steady state turns unstable and LC oscillations arise. The small panels on top show again the pair of complex eigenvalues with changing sign of the real part. In contrast to the supercritical HB, the arising LC is unstable which is indicated by blue circles. Additionally, the unstable LC can turn stable and therefore, also oscillations arising from subcritical HB are of interest in this work. Note that close to a subcritical HB, there can be regions in parameter space (in this case to the right of the HB) where a stable steady state and stable LC oscillations coexist.

2.2. Quantifying sensitivity: The chosen measures

In this work, a system characteristic is considered being robust if it is maintained when the system faces environmental changes (section 1.2). This work examines biological processes yielding limit cycle oscillations (section 2.1.6). The system characteristics of the oscillations which are examined are the period and the amplitude as visualized in Figure 1.1 and described in section 1.3. The environmental changes applied are represented via changes in the parameter values of the rate coefficients and nl-parameters of the underlying ODE system (section 2.1). Stoichiometry, cooperativity parameters, and the functional form of the rate equations remain unchanged.

In the following, an example for different robustness or sensitivities of the period of two oscillations is explained (Figure 2.5). Given is an ODE system that shows oscillatory behavior for two different sets of kinetic parameters (Figure 2.5 left, blue curve: parameter set 1, red curve: parameter set 2). For each parameter set, the solution oscillates with a certain period being characteristic for the parameter values. Here, it is $T_1 = 2.2$h and $T_2 = 0.3$h for parameter set 1 and 2, respectively. For the first parameter set holds that by increasing the value of the parameter k_1 by 10%, the period is increased to 3.3h. For the second parameter set an increase of the parameter k_1 by 10% leads to a decrease of the period to 0.06h. The question is for which parameter set the period of the oscillation is more sensitive. In Figure 2.5, one would intuitively expect the period at parameter set 2 to be more sensitive than the period at parameter set 2. However, because of the different time scales, the absolute period change for changing k_1 is smaller for parameter set 2 ($|\Delta T_2| = |0.06\text{h}-0.3\text{h}| = 0.24\text{h}$) than for parameter set 1 ($|\Delta T_1| = |3.3\text{h}-2.2\text{h}| = 1.1\text{h}$). In order to make robustness comparisons for oscillatory processes of different time scales possible (for example, circadian oscillations with a period of around a day opposed to calcium oscillations with periods of at most minutes), relative changes are going to be compared. For the first parameter set, the period length increases by 50% for the considered change in k_1. For the second parameter set, the same relative change in k_1 decreases the period by 80% of the original value. Hence, for the described case, the period of the oscillations at the first parameter set is less sensitive or more robust than the period of the oscillations at the second parameter set with respect to changes in k_1. In order to express the differences in robustness or sensitivity of processes of different time scales quantitatively, a measure for sensitivity is used which is based on the principle explained here.

This study employs a robustness measure which was introduced for the examination of steady states (Krüger and Heinrich [2004]). It is inspired by metabolic control analysis (MCA), where a response coefficient or control coefficient is defined by the relative effect on a system variable (flux or concentration) due to a rela-

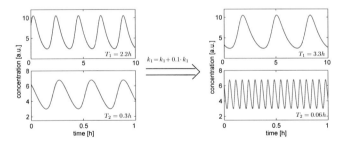

Figure 2.5.: Example for different period sensitivities. Sustained oscillation for two different parameter sets for the same ODE system are shown on the left. On the right, the oscillations for parameter k_1 increased by 10% are given after the transient phase. The period sensitivity with respect to changes in k_1 is higher for parameter set 2 (red) than for parameter set 1 (blue).

tive infinitesimal change of a parameter (Kacser and Burns [1973], Heinrich and Rapoport [1974]). A sensitivity measure for properties of oscillatory systems for infinitesimal changes has been proposed (Wilkins et al. [2009]). For simplicity, in this thesis, an adaptation of the measure from MCA to transient states as oscillations and oscillatory properties as period or amplitude using finite instead of infinitesimal changes is employed (Wolf et al. [2005]). The basis for the measure are the sensitivity coefficients R_{p_l} defined by

$$
\begin{aligned}
R_{p_l}^T &= \frac{\Delta T/T}{\Delta p_l/p_l} \quad \text{for the period and} \\
R_{p_l}^A &= \frac{\Delta A/A}{\Delta p_l/p_l} \quad \text{for the amplitude.}
\end{aligned}
\tag{2.5}
$$

The period T of an LC oscillation is mathematically defined in section 2.1.6. Recall that the amplitude of a species is the concentration difference from its largest concentration value and its lowest during one cycle. Since each of the analyzed oscillatory systems consists of more than one species, the arithmetic mean value of the amplitudes of all species, A, is examined as overall amplitude characteristic (discussed in section 2.5.2). In Equation 2.5, $\Delta T = T_l - T$ and $\Delta A = A_l - A$; T, A are the period and mean amplitude, respectively, for the unperturbed parameter set, and T_l and A_l the period and mean amplitude, respectively, for the parameter p_l being perturbed. Analogously, $\Delta p_l = p_l^{pert} - p_l$ is the difference between the perturbed (p_l^{pert}) and the unperturbed parameter value p_l.

The sensitivity coefficients in Equation 2.5 quantify relative changes in period and amplitude, $\Delta T/T$ and $\Delta A/A$, for relative changes in a single parameter p_l. Sensitivity coefficients take positive values if the direction of the applied parameter perturbation is the same as that of the resulting change in period or amplitude. They take negative values if the directions of parameter perturbation and change of the examined property are opposite. A sensitivity coefficient of zero indicates that the period or mean amplitude is not changing for the applied parameter change.

The period sensitivity coefficients in the example described above and given in Figure 2.5 are: $R_{k_1}^T = \frac{3.3h - 2.2h}{2.2h} \frac{k_1}{1.1 \cdot k_1 - k_1} = \frac{1.1}{2.2 \cdot 0.1} = 5$ for parameter set 1 and $R_{k_1}^T = \frac{0.06h - 0.3h}{0.3h} \frac{k_1}{1.1 \cdot k_1 - k_1} = \frac{-0.24}{0.3 \cdot 0.1} = -8$ for parameter set 2. Comparing the absolute values of the sensitivity coefficients leads to the same conclusion as above. The absolute sensitivity coefficient of the first parameter set ($|5|$) is smaller than that of the second parameter set ($|-8|$). Hence, the period of the ODE system at the first parameter set is more robust than the period of the ODE system at the second parameter set with respect to changes in k_1.

In order to obtain an average effect of the perturbation applied, the period sensitivity σ_T and amplitude sensitivity σ_A is defined by (in analogy to the measure introduced in Krüger and Heinrich [2004]):

$$
\begin{aligned}
\sigma_T &= \sqrt{\frac{1}{r}\sum_{l=1}^{r}(R_l^T)^2} \quad \text{for period sensitivity and} \\
\sigma_A &= \sqrt{\frac{1}{r}\sum_{l=1}^{r}(R_l^A)^2} \quad \text{for mean amplitude sensitivity}
\end{aligned}
\tag{2.6}
$$

at a specific parameter set for the ODE system, with r being the number of perturbed parameters. This is the quadratic mean over the sensitivity coefficients of all parameters. The squares are applied in order to avoid cancellation of single parameter effects. The measures σ_T and σ_A indicate how strongly perturbations of rate coefficients or nl-parameters influence the period and the amplitude, respectively. Since only ratios of concentrations, period and parameter values are used, σ_T and σ_A are dimensionless. They are normalized by the number of parameters to enable a comparison between the sensitivities of the oscillatory properties of ODE models yielding different numbers of parameters in their ODE system. σ_T and σ_A only take non-negative values. Low values of the sensitivity measures determine low sensitivity and hence high robustness, high values indicate high sensitivity and therefore low robustness.

In this work, in order to keep the calculation effort small and the approach simple, the perturbation applied to the single parameters to calculate their sensitivity coefficients is an increase by 2% (discussed in section 2.5.4). If not stated differently, the overall sensitivities σ_T and σ_A are formed by the quadratic mean over the sensitivity coefficients of all rate coefficients and nl-parameters (discussed in section 2.5.4).

2.3. Sensitivity estimation

The aim of this analysis is to compare the period and amplitude sensitivities of different biological processes. To this purpose, published ODE models for the processes in question are examined. In literature, the ODE models are published together with certain numerical values for the incorporated kinetic parameters of the ODE system, the so-called reference parameter set.

The assumptions of the model on the structure of the biological process are encoded in the interactions between the species which are completely given by the ODE system, its stoichiometric matrix and the form of the rate functions, even without particular values for the kinetic parameters. Some further structural assumptions on the cooperativity of reactions is implemented by the values of the cooperativity parameters. Therefore, in this work, the species and interactions between species given by the ODE system are adopted unchanged from the literature, as well as the values of the cooperativity parameters (as long as not stated differently), throughout the sensitivity analysis.

In contrast, the other kinetic parameters as rate coefficients and nl-parameters do not contribute structural information. As long as these parameters keep their (positive) sign, all positive and negative interactions of the process are kept as in the published model. Additionally, the kinetic parameters can hardly be measured (Klipp and Liebermeister [2006], Gerard et al. [2009]). These insecurities in the actual values of the kinetic parameters are taken into account in the sensitivity estimation process by not only examining the ODE system at the reference parameter set. Instead, in order to avoid further assumptions, for each model, this work analyzes the sensitivities of period and amplitude for 2500 different, randomly chosen sets of parameters (with fixed cooperativity parameters) giving rise to sustained oscillations. This approach allows to differentiate whether particular sensitivity properties are due to structural assumptions on the biological process or arise solely from the choice of the values for a particular parameter set.

For the computations, Matlab (Matlab 7.11, The MathWorks, Inc.) is used. The analysis of the sensitivity of the oscillatory properties of the ODE systems is carried out as depicted in Figure 2.6 for the example of the calcium oscillation model by Goldbeter et al. [1990]. The details of the work-flow are explained in this section.

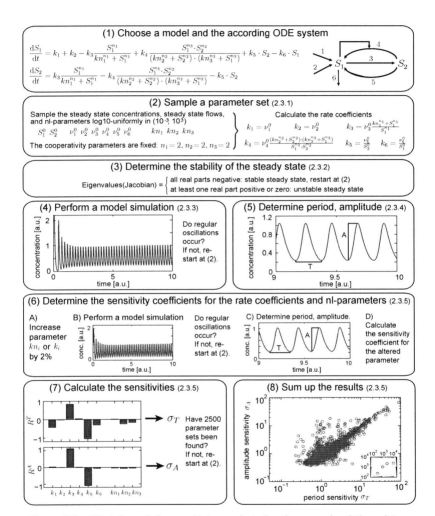

Figure 2.6.: Work-flow of the sensitivity analysis for the example of the calcium oscillation model of Goldbeter et al. [1990]. The sensitivity analysis can be subdivided into eight parts. The number in brackets behind the name of each part gives the section where details on the respective part can be found.

2.3.1. Parameter sampling

In this work, the examined biological processes are represented by published ODE models. The values of the nl-parameters and rate coefficients of the models are sampled in order not to put further assumptions in the analysis and to discriminate effects of the structure from effects of the choice of the parameter values.

The ODE models examined in this work contain from nine to 46 nl-parameters and rate coefficients. This makes a systematic covering of the parameter space infeasible

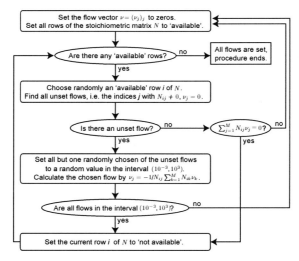

Figure 2.7.: Illustration of the sampling of the steady state flow vector $\nu^0 = \nu$. Basis is the condition $N \cdot \nu = 0$ for ν being the flow vector at steady state for the stoichiometric $m \times M$-matrix N. The order in which the flows are set has to be chosen randomly in order not to impose bias upon the sampling results. This procedure is explained in the decision diagram. For accelerating the sampling process, independent parts of the stoichiometric matrix N and according sets of flows can be proceeded separately. Alterations needed to determine r steady state concentrations subject to q conservation relations (point (ii)): The initial vector ν has to be replaced by $\nu_S = (S_1^{tot}, \ldots, S_q^{tot}, 0, \ldots, 0) \in \mathbb{R}^{(r+q)}$, where S_j^{tot} is the total concentration of the jth conservation relation. The stoichiometric matrix N has to be replaced by a $(q \times (q+r))$-matrix N_S such that $N_S \cdot \nu_S = 0$ delivers the conservation relations.

and necessitates a random sampling approach. The classical strategy for parameter sampling would be to sample the nl-parameters and rate coefficients randomly (as done in e.g. Stelling et al. [2004a], Tsai et al. [2008], Hafner et al. [2009]). This approach requires knowledge of the sampling intervals of the kinetic parameters which are hardly known (e.g. Klipp and Liebermeister [2006], Gerard et al. [2009]). Instead, this work employs a bottom-up approach of sampling in which the steady state concentrations, steady state flows and nl-parameters are sampled and the rate coefficients are calculated from those quantities (Steuer [2007]). Therein, no assumptions have to be made on the actual ranges for the kinetic parameters, but only on the steady state species concentrations, steady state flows and nl-parameters.

Additionally, since sustained oscillations shall be observed, only sampled parameter sets yielding a biologically feasible, i.e. non-negative, unstable steady state are further examined to reduce the calculation effort (see also section 2.3.2). However, in the classical sampling approach, this would require solving a set of non-linear equations to determine the steady state. Then, it might occur that for a given system and a particular set of values for the parameters, no biologically feasible steady state exists. Another extreme would be multi-stationarity, i.e. the occurrence of multiple positive real steady states where the choice for the steady state to be examined is not trivial. These issues are circumvented using the bottom-up sampling approach in this work. Since the steady state concentrations are directly sampled, they do not have to be calculated and also the issues concerning the number or choice of the steady states are avoided. Details of the sampling procedure are described in the following.

(i) A feasible set of steady state reaction rates (steady state flows) ν^0 satisfying the steady state relation $N \cdot \nu^0 = 0$ is sampled randomly in the interval $(10^{-3}, 10^3)$. Thereby, a random distribution is used which keeps the decadic logarithms uniformly distributed, hereafter referred to as log10-uniform distribution. Using this distribution, each of the six orders of magnitude is represented with the same probability. The log10-uniform distribution is used in order to cope with the fact that different flows and species concentrations may occur at very different orders of magnitude. The technical details of the procedure of sampling the steady state flows are illustrated in Figure 2.7.

(ii) The steady state concentrations are sampled in the interval $(10^{-3}, 10^3)$ using a log10-uniform random distribution. Concentrations of species denoting probabilities occurring in two of the calcium models (De Young and Keizer [1992], Sneyd et al. [2004], section 7.2) are not allowed to exceed a value of 1. For these species, the sampling interval is adapted to $(10^{-3}, 1)$. If conservation relations exist in the ODE system, the conserved total concentrations are sampled log10-uniformly in the interval $(10^{-3}, 10^3)$ and the steady state concentrations contributing to the conservation relation are set accordingly in a random manner. Thereby, the procedure for sampling the steady state flow vector is used (see Figure 2.7, necessary changes are described in the legend).

(iii) All nl-parameters are sampled log10-uniformly in the interval $(10^{-3}, 10^3)$.

(iv) From the fixed cooperativity parameters and the sampled steady state concentrations, steady state flows, and nl-parameters, the rate coefficients can be uniquely determined via calculation.

Note that this sampling approach ensures a specific probability distribution for all possible combinations of steady state concentrations of the species, steady state flows and nl-parameters. However, it does not necessarily imply that the obtained rate coefficients obey the same distribution or are even within a specific region. The choice of the sampling interval and the general sampling approach is discussed in section 2.5.4.

2.3.2. Determining the stability of the steady state and performing model simulations

After sampling a set of kinetic parameters and steady state concentrations as explained in section 2.3.1, the stability of the steady state can be determined with the help of the eigenvalues of the Jacobian matrix (see section 2.1.5). In general, unstable steady states provide the possibility of the occurrence of oscillations. However, close to a subcritical HB, oscillations can also occur while experiencing a stable steady state (see section 2.1.7). In this work, the ODE system is only further analyzed for the particular sampled parameter set if the sampled steady state is unstable. It is neither detected nor focused on which particular bifurcation type is responsible for the instability of the steady state.

With a parameter set that yields an unstable steady state, the ODE system is solved numerically. Initial values $S(0)$ are chosen to be 95% of the steady state concentration values S^0: $S(0) = 0.95 \cdot S^0$. In case of conservation relations, all but one of the contributing species are initially set to 95% of their steady state concentrations, the remaining initial species concentration is set accordingly to retrieve the conserved value. The system is integrated with an explicit one-step Runge-Kutta solver (ode45 in Matlab 7.11, The MathWorks, Inc.) or a variable-order solver based on the numerical differentiation formulas (ode15s in Matlab 7.11, The MathWorks, Inc.). It depends on their initial performance, i.e. which solver reaches the later integration time point during the first 2 seconds, whether the solver ode45 or ode15s is used. The ODE system is integrated up to a time variable of 20000 unless regular oscillations (see section 2.3.3) are detected earlier. Error tolerances for the integration are set to 10^{-6} and 10^{-8} for relative and absolute errors, respectively. The integration is interrupted if it takes longer than three minutes for ODE systems with less than or exactly ten variables or longer than five minutes for ODE systems with more than ten variables.

2.3.3. Checking for regular oscillations and determining period and amplitude

In this study, only regular oscillations as defined in the following are considered. A solution of the ODE system is classified as regular oscillation if the species with the largest amplitude shows five consecutive equal maxima with equal inter-spike intervals. The maxima and inter-spike intervals are considered to be equal if they differ by less than 10^{-3}% and 0.1%, respectively. Period and amplitude are the characteristics of the oscillation that will be examined throughout the thesis, see Figure 1.1. In analogy to the definition given in section 1.3, the period of the oscillation is measured as the inter-spike interval and the amplitude of a species is determined as the difference between the global maximal and the global minimal value during one cycle. If the period and amplitudes of the regular oscillations have been determined, the system is integrated again from the end-point of the previous calculation with 100-fold less tolerances up to 100-fold of the calculated period. The period and amplitudes of this solution are determined. If the oscillatory properties agree for both simulations with different calculation tolerances, the parameter set is further investigated; if not, it is discarded.

2.3.4. Calculation of the sensitivity

In order to estimate the period and amplitude sensitivity as described in section 2.2, the degree of alteration the oscillatory properties induced by parameter perturbations are examined. Each rate coefficient and nl-parameter is increased separately by 2% and the system is integrated again as described in section 2.3.2. The end-point from the computation without parameter perturbation is taken as initial values. The period and amplitude of the oscillation are determined as described in section 2.3.3. Then, for each parameter, the period and amplitude sensitivity coefficient can be calculated as given in Equations 2.5 using the period and amplitude of the unperturbed system. If for a certain parameter perturbation no regular oscillations are found, the parameter set is discarded. If the sensitivity coefficients for all rate coefficients and nl-parameters have been determined, the period and amplitude sensitivities σ_T and σ_A are calculated as indicated in Equations 2.6. If not stated differently, for each model it is sampled until the sensitivities σ_T and σ_A have been determined for 2500 parameter sets. This includes examination up to millions of parameter sets. The period sensitivity and amplitude sensitivity data of a sensitivity analysis are summarized in a scatter plot. The resulting data cloud displays the robustness potential of the model structure, i.e. it gives an impression of the general tendency for the period and amplitude robustness and of possible and probable combinations of period and amplitude robustness. It is further investigated with statistical methods described in section 2.4.

2.3.5. Oscillation probability

The number of parameter sets which are sampled in total in order to obtain 2500 parameter sets of which the period and amplitude sensitivity can be calculated is obtained automatically during the analysis for each model. This number gives a hint on the percentage of found oscillations, or on the size of the oscillatory region. Although this type of robustness of oscillations is not of central interest in this work, it is indicated together with the characteristics of the period and amplitude sensitivity distributions.

2.4. Methods of statistical analysis

In this work, sensitivity analyses are performed for many different models. For each model, the analysis generates large amounts of data, having a number of 2500 values for period sensitivity and amplitude sensitivity each. Therefore, methods of descriptive and comparative statistics are essential. Furthermore, since this work is based on a Monte-Carlo approach for parameter sampling and the amount of sensitivity data obtained is large but still not exhaustive for any ODE system examined, inferential statistics are needed. All statistical methods employed in this work are described in this section. The information provided here and further information can be found in textbooks for statistics, e.g. Boslaugh and Watters [2008], Rinne [2008], Gibbons and Chakraborti [2011].

2.4.1. Median, quartiles, 90% data range

In order to specify central tendencies and dispersion, descriptive statistics are used. In fact, the data obtained by the sensitivity analyses reveal to follow skewed, heavy-tailed distributions with a large number of sensitivity values which are very far away from the central values. Therefore, it is not appropriate to use the arithmetic mean value and related measures (discussed in section 2.5.2). Instead, this work employs measures of special percentiles and percentile distances for describing the central tendency or dispersion of the data.

Definition of percentiles

For the definition of a percentile, the order of the sample data points is considered. Let $Y = \{y_j\}$ be a sample of n elements, i.e. data points, and assume without loss of generality ordering, i.e. $y_1 < y_2 < \ldots < y_{n-1} < y_n$. In general, the pth percentile,

Q_p, is defined to be the value such that at most a proportion of $p\%$ of the data points of the sample has smaller and at most a proportion of $(100 - p)\%$ has larger values than the Q_p. This means Q_p is a value between the lowest $p\%$ and the largest $(100 - p)\%$ of the sample. In this work, it is defined by

$$Q_p = \begin{cases} \frac{y_{n \cdot \frac{p}{100}} + y_{n \cdot \frac{p}{100} + 1}}{2} & \text{for } n \cdot \frac{p}{100} \text{ integer} \\ y_{round(n \cdot \frac{p}{100} + 0.5)} & \text{else.} \end{cases} \tag{2.7}$$

Thereby, $round(n \cdot \frac{p}{100} + 0.5)$ defines the lowest integer being larger than $n \cdot \frac{p}{100}$. The different cases are necessary in order to determine a unique value for the percentiles. Also other definitions of percentiles can be found, however, their values converge for the number of points in the data set getting large.

Definition of special percentiles and percentile distances

The most frequently used percentile is the 50th percentile, also called the median with symbol μ. At most 50% of the data points in the sample are smaller and at most 50% are larger than the median. If n is uneven, the median is given by $y_{(n+1)/2}$. If n is even, any value between $y_{n/2}$ and $y_{n/2+1}$ satisfies the properties of the median. The definition from Equation 2.7 sets the value of the median to $(y_{n/2} + y_{n/2+1})/2$.

Two other important percentiles are the 25th and the 75th percentiles which are also known as the first and third quartile, respectively. As their names state, the quartiles indicate the division of the data in quarters. In the interval formed by the first and third quartile, the middle 50% of the data points are located.

The typical measure of dispersion or variability of the data is the standard deviation. However, it relies on the distance of the data values from the arithmetic mean and therefore has only minor descriptive power in the present case of skewed and heavy-tailed data distributions. Instead of the standard deviation, the range is often taken as a non-parametric measure without any assumptions on the underlying data distribution. The range is the interval or the length of the interval given by the smallest and the largest element of the data set. The range is very sensitive to extreme values which are probably not similar when performing a second sampling. Therefore, another measure of dispersion is used in this work: The interval given by the fifth and the 95th percentiles. Since it captures the range in which 90% of the data is located, its length is hereafter referred to as 90% data range (90%R).

2.4.2. Confidence interval of the median

A Monte-Carlo approach as applied in this work includes the fact that not every possible of the infinite number of parameter sets for an ODE system is examined.The data analyzed here is considered to be only a sample from a whole population whose statistical properties are unknown. In order to conclude from properties of the sample to general properties of the model, and especially to assess how sure one is about this conclusion, methods of inferential statistics are used. One of them is the calculation of the confidence interval of the median which determines how exact the median of the sample matches the true median of the population.

In detail, this work employs the 95% confidence interval (CI) of the median. The 95% CI of the median is an interval enclosing the calculated median of the sample. It is the region in which 95% of the sample medians would be found if doing new samplings. Therefore, one could infer that with a probability of 95% also the actual median of the population can be found in this interval.

In order to calculate CIs of the median, an equation introduced in the 1930s is used (Thompson [1936]). In short, the necessary mathematics read as follows: Let $Y = \{y_j\}$ be a sample of n elements from a population whose distribution characteristics are unknown and shall be inferred from the sample. Without loss of generality, one can assume that the elements of the sample are sorted in increasing order. y_k is defined as the kth element. If μ_Y is the actual median of the distribution (which cannot be directly calculated from the sample), the symmetric confidence interval around the median μ_Y can be retrieved using the relation

$$P(y_k < \mu_Y < y_{n-k+1}) = 1 - 2 \cdot \left(\frac{1}{2^n}\right) \sum_{i=0}^{k-1} \binom{n}{i} \quad \text{for } 2k < n+1. \quad (2.8)$$

This formula is consistent with the general formula for estimates of confidence intervals of general quantiles (Gibbons and Chakraborti [2011]). The equation gives the probability of the true median μ_Y of the population from which the sample was drawn to lie in the interval (y_k, y_{n-k+1}) which is completely determined by the index

Table 2.1.: Indices determining the 95% CIs for median values in dependence on the number of sets in the sample, n, calculated based on Equation 2.8. The intervals are formed by the kth and $(n - k + 1)$st element of the ordered sample. The exact confidences of the intervals are given (in %).

n	10	50	100	500	1000	1500	2000	2500
k	2	18	40	228	469	712	956	1201
confidence	97.85	96.72	96.48	95.59	95.37	95.32	95.34	95.23

k. In order to determine the 95% CI, one has to determine the index k of the sample such that the right hand side of Equation 2.8 takes a value of 0.95. Since the exact value is hardly reached, in general the k reaching the next largest value greater than 0.95 is considered. Calculations for a 95% CI lead to explicit values for the indices k in dependence on the sample size n given in Table 2.1. For the index being $k - 1$, the confidence would lie below 95%. In this work, the sample size is often $n = 2500$ since this is the number of parameter sets yielding regular oscillations until which it is sampled. Be aware that the CI is only symmetric with respect to the index. In most cases, the values for y_k and y_{n-k+1} do not have the same distance from the median value.

2.4.3. Box-plot visualization

The major properties of a one-dimensional data sample or data set can be summed up in so-called box- or box-and-whiskers plots. Figure 2.8 shows an example of a typical box-plot as employed in this work for a data set of 100 data points obtained in a sensitivity analysis. The red line in the middle of the box indicates the median.

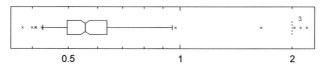

Figure 2.8.: Visualization of data by a box-plot on log-scale. Shown are the median (red line), the 95% CI of the median (notch), the range between Q_{25} and Q_{75} (box), the 90%R (whiskers), the data values outside the 90%R as crosses with their actual values or not using the actual values (equidistantly behind the dotted line, their number is indicated).

The notch shows where the 95% CI of the median lies. The box begins at the first quartile and ends at the third quartile. The tips of the whiskers indicate the fifth and the 95th percentiles, their difference indicates the 90%R.

The remaining 10% of the data points of the data set, the smallest and the largest 5%, are marked in the plot by crosses. Most of them are thereby positioned according to their actual value. In the example in Figure 2.8, this is the case for all five lowest data values and two of the five largest data values. Plotting the exact value of data points whose values are far from the central tendency sometimes makes it difficult or impossible to analyze the rest of the data, even if already plotting on log-scale as in the example, where the three largest data values are comparably high taking values of 3.7, 16.9 and 441.6. These data points are separated from the rest of the plot by a dashed line and not plotted according to their values but only equidistantly from each other. The number of these data points is given close to their plotting region. In this case (Figure 2.8), there are three data points not plotted according to their actual data values due to visualization needs. Be aware that the definition of the whiskers and also of the crosses differ from others employed in literature (Boslaugh and Watters [2008], Rinne [2008]) in order to visualize the properties of the heavy-tailed and skewed data distributions better.

2.4.4. Spearman's rank correlation coefficient R_S

In the course of the sensitivity analysis, both the period sensitivity and the amplitude sensitivity are determined for each sampled parameter set for the ODE system. Hence, period and amplitude sensitivity are dependent samples. The correlation between them is useful to assess to which degree the two sensitivities are related. For this purpose Spearman's rank correlation coefficient R_S is calculated. In contrast to other correlation measures as Pearson's correlation coefficient, R_S has the advantage to be able to detect also nonlinear correlations. Additionally, it is insensitive to extreme data values and to the particular shape of the distributions of the compared samples as it is based on rankings within the data samples, and not on the actual data values.

In general, assume $(X, Y) = \{(x_i, y_i)\}$ to be paired samples with n pairs. For calculating R_S, it is necessary to determine the order of the x_i and the y_i. Define

$rx_i = rank(x_i)$ and $ry_i = rank(y_i)$ the ranks of the separated samples X and Y in ascending order. This means that $rx_i, ry_i \in \{1, 2, \ldots, n-1, n\}$ are integer numbers between one and n and if $x_i < x_j$ also $rx_i < rx_j$, and if $y_i < y_j$ also $ry_i < ry_j$. Spearman's rank correlation coefficient is calculated by

$$R_S = 1 - \frac{6 \sum_{i=1}^{n} (rx_i - ry_i)^2}{n(n^2 - 1)} \tag{2.9}$$

in accordance with the definition in Boslaugh and Watters [2008]. R_S takes values from the interval $[-1, 1]$. Thereby, $R_S = 1$ means perfect concordance and is achieved if the order of the sample X is exactly the same as that for the sample Y. $R_S = -1$ means perfect discordance where the order in Y is the reversed order of X. The closer R_S is to zero, the less the data are correlated, no correlation appears for $R_S = 0$. The significance of the correlation can be also assessed (section 2.4.5). In this work, the general rules for interpreting intermediate values of R_S are used (Boslaugh and Watters [2008]):

- $0.9 \leq |R_S| \leq 1$ indicates a very strong correlation

- $0.7 \leq |R_S| \leq 0.9$ indicates a strong correlation

- $0.5 \leq |R_S| \leq 0.7$ indicates a moderate correlation

- $|R_S| < 0.5$ indicates low correlation.

2.4.5. Testing the significance of differences

The principle of hypothesis testing is very useful in comparative statistics. It is employed in this work for determining the significance of differences of the sensitivity distributions of different models. The procedure can also be used if only testing hypotheses of single samples.

Principle of hypothesis testing, p-values

(i) A null hypothesis H_0 is stated about a specific relationship between the distributions of the populations from which the samples are taken. The aim of the test is to determine a handle whether to reject or to accept this hypothesis for the population distributions on the basis of the samples.

(ii) An alternative hypothesis H_1 is stated which is inconsistent with the null hypothesis. The alternative hypothesis is accepted if the null hypothesis is rejected.

(iii) The value of the test statistic is calculated. The test statistic is defined such that its distribution is known given that the null hypothesis is correct. Often, also assumptions on the distribution of the population from which the samples are drawn are made in order to determine the shape of the test statistic distribution under the null hypothesis.

(iv) A p-value is calculated which indicates the probability to obtain the actual value of the test statistic (or one more extreme in disfavor of the null hypothesis) given that the null hypothesis is true. The p-value indicates at which level the null hypothesis is rejected. Typical rejection levels are 5% or 1%. In this work, a rejection level of 1% is chosen, i.e. the null hypothesis is rejected for p-values smaller than 0.01.

Significance of Spearman's rank correlation coefficient R_S

The significance of Spearman's rank correlation coefficient R_S is assessed with the test proposed in Boslaugh and Watters [2008]. Suppose a sample with n pairs of observations (X_i, Y_i). Let r_s be Spearman's rank correlation between the observed entity X and the observed entity Y. For testing the significance of the obtained correlation, the null hypothesis H_0 is tested of the true rank correlation R_S between the random variables X and Y from which only an n-sample was drawn to determine the observed correlation r_s is zero, $R_S = 0$. The alternative hypothesis H_1 is that $R_S \neq 0$. The test statistic is the standard normal distribution, and if the null hypothesis is true, the value of the test statistic is zero. For an n-sample, the value of the test statistic is calculated by $z = r_s \cdot \sqrt{n-1}$. The p-value for rejecting the hypothesis is then determined by estimating how probable it is to obtain the value z or more extreme values by chance given that the z-values are standard-normally distributed.

For example, in case it has been determined that $r_s = 0.5$ in a sample with 500 parameter sets, the p-value is the probability that the standard normal distribution takes values of $z = 0.5 \cdot \sqrt{500 - 1} = 11.17$ or higher. Accordingly, the p-value in this case is $2.9 \cdot 10^{-29}$, and the null hypothesis of $R_S = 0$ can be rejected, the positive correlation between X and Y is significant.

Note that the significance of an observed correlation depends on size of the sample. If the same $r_s = 0.5$ had been observed for a sample of size $n = 15$, this delivers a p-value of $p = 0.0307$, and thus the null hypothesis could not be rejected at a 1%-confidence level. Contrariwise, for a sample size of $n = 2500$ as frequently encountered in this thesis, already extremely weak correlations of $|r_s| = 0.0466$ are considered to be significant at the 1%-confidence level. Consequently, in this thesis, the significance of correlations, i.e. the p-values, are only provided in case of smaller sample sizes.

The Mann-Whitney-U test

In the form as presented here (adapted from Gibbons and Chakraborti [2011]), the Mann-Whitney-U (MWU) test, also called Wilcoxon rank sum test, is a very general test for comparing two distributions of populations from which samples X of size n_X and Y of size n_Y are taken. This test is non-parametric and has the advantage of not relying on any assumptions of specific shapes of the compared distributions. It enables an unbiased comparison of the results of the sensitivity analysis which do not follow a normal, symmetric or any other specific distribution.

In order to be in accordance with the general use of rejection of the null hypothesis for low p-values, the one-sided MWU test is used. Therefore, without loss of generality, assume that for the samples X and Y, the median of X is smaller than the median of Y. For simplicity, the same notation as for the samples is used for the distributions, X and Y, respectively.

(i) The null hypothesis H_0 is that the distributions of the compared populations are identical. Expressed mathematically with the probability functions P_X and P_Y of the underlying distributions, this is given by $P_X(X < x) = P_Y(Y < x)$ for all real values x.

(ii) The alternative hypothesis H_1 is that the distribution of the population from which sample X was taken is stochastically smaller than the population from which sample Y was taken, $X <_{ST} Y$. Mathematically, being stochastically smaller is

Table 2.2.: Table for the results of the MWU test. The identifiers (id_X and id_Y) of the models in the order of comparison are given, the oscillatory characteristic of which the sensitivity is examined (period, per., or amplitude, ampl.) is indicated. The p-value (MWU p) is the most important result of the test. n_X and n_Y are the numbers of elements of the samples, U the value for the calculated test statistics. z-val gives the value in the standard normal distribution which complies with the calculated U value in the normal distribution with standard deviation and mean as described in the text.

id_X		id_Y	sensitivity	MWU p	n_X	n_Y	U	z-val
model A	⇔	model B	per.	$3.4 \cdot 10^{-3}$	2500	2500	6101599	-2.93

defined by $P_X(X < x) \geq P_Y(Y < x)$ for all x with a true inequality $P_X(X < x) > P_Y(Y < x)$ for at least one x (Gibbons and Chakraborti [2011]). This means that if taking randomly one data value from each of the two compared distributions, it is more probable that the value drawn from X is smaller than the value drawn from Y than the other way round. Being stochastically larger is defined analogously.

(iii) The test statistic of the MWU test relies on the ranking of the combined sample $X \cup Y$. The MWU test statistic is given by

$$U = \sum_{i=1}^{n_X} \sum_{j=1}^{n_Y} S(x_i, y_j), \text{ where } S(x_i, y_j) = \begin{cases} 0 & \text{for } x_i \leq y_j \\ 1 & \text{for } x_i > y_j \end{cases} \tag{2.10}$$

as in Gibbons and Chakraborti [2011]. Every possible pair of one data value in sample X, x_i, and one in sample Y, y_j, is looked at and it is counted how often the value x_i exceeds the value y_j. The less often this event occurs, i.e. the smaller U is, the more probable it is that $X <_{ST} Y$, and consequently, the less probable it is that X and Y have the same distribution.

(iv) If the null hypothesis of similar distributions of X and Y is true, the test statistic is approximately distributed as the normal distribution with mean $n_X n_Y / 2$ and standard deviation $\sqrt{n_X n_Y (n_X + n_Y + 1)/12}$ for large sample size ($min(n_X, n_Y) > 30$). Hence, for the calculated value of U in the normal distribution with mean and standard deviation as given above, it can be derived the according z-value from the standard normal distribution. The p-value, which is the probability of the occurrence of the obtained value for U or lower values, can be approximated. It is $P(x < \text{z-value})$ for the probability function P of the standard normal distribution.

For p-values smaller than 0.01, the null hypothesis of X and Y having the same distribution is rejected. For smaller sample sizes, tables of critical values delivering which values of U are necessary to reject the null hypothesis at level 0.05 or 0.01 are available (e.g. Rinne [2008]).

The results of MWU tests in this work will be provided similarly to the presentation of the information in Table 2.2.

2.5. Summary and discussion: Robustness measure and methods

2.5.1. ODE models and oscillation detection

For the analysis of the sensitivity of period and amplitude of oscillatory processes, published ODE models are used. The reason for using ODE models instead of other model approaches as network analysis, stoichiometric analysis and structural kinetic modeling is that the latter do not deliver enough detail to examine these properties of oscillations (Steuer [2007]). This work focuses on the analysis of characteristics of sustained oscillations. They are considered to be represented by stable limit cycle oscillations in the ODE models in order to have a unique period and amplitude (section 2.1.6).

Numerically, the oscillations are detected as solutions of the ODE systems showing regular oscillations, because for simulations in a finite time and including calculation errors, it cannot be proven that the oscillations are actually residing on a limit cycle (section 2.3.3). In the whole analysis, parameter sets where the oscillations are extremely slowly converging to the LC or where the calculation needs too much time are not considered as "oscillating". Also, the sensitivities of periodic behavior yielding two (or more) exactly identical peak values with different interspike intervals are not further examined. Thereby, one might lose some oscillations which would in principle exist in nature. However, some numerical, threshold and time limits need to be fixed in order to automatize the procedure of detection of oscillations and to be able to examine large numbers of parameter sets. In addition, due to numerical limitations, neither oscillations with a very low amplitude in all species nor oscillations with very low and very high periods are considered in the analysis. This is only a minor restriction since those types of oscillations would be hardly detectable in biological experiments and may be indistinguishable from other types of transient behavior in the cell or organism.

Of course, other possibilities of analyzing oscillations *per se* exist, e.g. one could characterize the onset behavior of oscillations, or the time until an LC or regular oscillation is reached. However, this is mainly of interest for oscillations which are switched on upon a stimulus and where the system resides on a steady state before. Additionally, if examining characteristics for other than limit cycle oscillations (or its numerical representation by regular oscillations), no unique period or amplitude can be determined and hence a sensitivity analysis would be biased by the choice of the peak to represent the amplitude and the choice of the inter-spike interval for the period.

Bifurcation analyses of particular parameter sets of particular models are introduced as means to examine potential sources of increased or decreased sensitivities (section 2.1.7). However, it is not the focus of this work to determine the bifurcation type from which the oscillations arise or to restrict the analysis to oscillations arising from certain types of bifurcations.

2.5.2. Mean amplitude as amplitude characteristic

In the course of the examination, the amplitude of a species signifies the distance between its lowest value and its highest during one cycle of the LC oscillation. Having usually more than one species for each system and not knowing which species is the interesting one, the amplitude property of an oscillation is chosen to be captured by the arithmetic mean value of the amplitudes of all species of the model. The sensitivity of the amplitude is hence based on changes in the mean amplitude (see section 2.2). For any oscillatory process with positive species concentration values, this ensures that the observed amplitude property is positive even if some of the species concentrations are constant and therefore yield a zero amplitude. Consequently, also the measure for amplitude sensitivity (Equations 2.5, 2.6) is well-defined.

By taking the mean amplitude, small changes due to parameter perturbation in species with high amplitudes might mask high changes occurring for species with lower amplitudes. However, the species yielding the largest amplitudes are those which might be detected most easily in a biological system and might serve as markers for the oscillatory activity. Therefore, changes in the mean amplitude are probably quite relevant for biological applications. Additionally, similar changes of the amplitudes of different species in opposite directions result in very low amplitude sensitivities using the mean amplitude although each single species might be heavily affected. Therefore, conclusions on low amplitude sensitivity need to be interpreted with caution. In contrast, if experiencing high amplitude sensitivities for the mean amplitude as amplitude characteristic, it can be concluded that at least some of the single species amplitude sensitivities are large. Altogether, if one is interested in the sensitivity of the amplitude of a particular species participating in a process, one can and should adapt the examination accordingly. Relative amplitudes (i.e. amplitude/maximal value) or maximal values may be also taken into account as amplitude characteristic. However, they constitute other amplitude characteristics of oscillations and may therefore lead to different results.

2.5.3. Sensitivity measure

The robustness of the period and amplitude are determined with sensitivity measures derived from MCA. These measures consider relative changes of period and amplitude for relative changes in parameters (section 2.2). This enables the comparison of processes of different orders of magnitude of species concentration or time scales as e.g. circadian oscillations with periods in the range of 24h and calcium oscillations with periods in the range of seconds to minutes. The period sensitivity and amplitude sensitivity are calculated as average of the sensitivities towards changes in the nl-parameters and rate coefficients of the ODE models (section 2.1.2). The characteristics of the mechanistic interactions between metabolites are not varied, meaning that stoichiometry, cooperativity parameters, and the functional form of the rate equations remain unchanged. These features, in contrast to the rate coefficients and nl-parameters, can be assumed to remain unaffected by perturbations such as agonist stimulation or temperature variations.

In the approach in this work, the sensitivity is calculated as quadratic mean over the sensitivity coefficients of the nl-parameters and the rate coefficients of the models. Other sensitivity measures are of course possible. In the Appendix, two examples of other sensitivity measures based on sensitivity coefficients are used for two models (section A.1.2). For example, if only considering the largest three sensitivity coefficients, the sensitivity distributions are considerably altered and strongly varied outcomes of the sensitivity analysis when comparing models may be obtained. In contrast, taking the quadratic mean only over the rate coefficients and leaving out the nl-parameters yields similar although not exactly the same sensitivity distributions.

This examination reinforces the observation from e.g. Stelling et al. [2004a] that sensitivity or robustness needs to be clearly defined, especially also the perturbation against which the sensitivity is assessed. Even stating that parameters shall be perturbed is not sufficient and can lead to more or less different sensitivity results for different sensitivity measures considering different parameters. In this work, the average over the sensitivity coefficients of all kinetic parameters is employed because it is the most general approach without favoring parameters or biasing the results if the sensitivities are asymmetrically distributed among the parameters. If one is interested in the impact of particular parameters on the period or amplitude, their sensitivity coefficients can be examined directly (as e.g. in sections 4.4.1, 5.3).

2.5.4. Sampling approach and sensitivity estimation

In order to distinguish between the effect of the choice of the particular parameter set and of structural properties on the period and amplitude sensitivity, the sensitivity is determined for a large number of parameter sets for each model. For obtaining the parameter sets, a Monte Carlo random sampling approach is chosen since a systematic examination of the parameter space is not feasible for the models yielding many kinetic parameters (from nine to 46 in this work). The random sampling approach (section 2.3.1) employs a log10-uniform distribution in order to account for the occurrence of different orders of magnitudes in species concentrations or flows. The steady state species concentration, the steady state flows and the nl-parameters are sampled in the interval $(10^{-3}, 10^3)$. The resulting unique set of rate coefficients is calculated from the sampled quantities. This bottom-up sampling approach (described in Steuer [2007]) is chosen in this work because it circumvents problems arising with choosing a steady state in case of multi-stationarity. In addition, ranges for kinetic parameters are hardly known (Klipp and Liebermeister [2006], Gerard et al. [2009]).

One possible issue in the sampling approach is this particular choice of the sampling region to the interval $(10^{-3}, 10^3)$. However, as being interested in the relative changes of the oscillatory properties due to small relative perturbations of the parameters, the results are quasi-independent of the choice of the sampling region (Appendix, section A.1.3). For example, a change in the sampling region for the flows corresponds to a rescaling in time. On average, this could lead to different periods, but the sensitivity measures remain unaffected. A change of the sampling interval of the concentrations and nl-parameters corresponds to a rescaling of the concentration unit, but the obtained sensitivity distribution does not change. Therefore, one can conclude that general results derived from the sensitivity analyses in this work are, if at all, only slightly influenced by the particular choice of the sampling interval.

Although the choice of the specific sampling interval reveals to only slightly influence the obtained sensitivity results, the sampling procedure still imposes a constraint to the parameter space being observed: The sampling intervals for steady state concentrations and nl-parameters are the same. This restriction is biologically plausible since nl-parameters mainly characterize the affinity of an enzyme or channel for its substrate or regulatory species. Thus, the sampling approach with a log10-uniform distribution over six orders of magnitude yields the possibility of examining low, intermediate and high affinities compared to the species steady state concentration. An example of which specific values occur when sampling a Michaelis-Menten term with the approach in this work is given in the Appendix (section A.1.4).

A sampled parameter set is discarded in the sensitivity analysis if the sampled steady state is stable or if for the original parameter set or any parameter perturbation no regular oscillations are found (sections 2.3.2, 2.3.3). In this work, the perturbation applied to the parameters to calculate the sensitivity coefficients is a relative increase by 2%. Since the parameters are obtained via a random sampling approach, the results are indeed approximately the same if applying a decrease of 2% (Appendix, section A.1.1). Additionally, since the sensitivity is calculated from relative changes, other sizes of parameter perturbations as 1% and 5% lead to similar sensitivity results (section A.1.1). For a parameter perturbation of 10%, some of the sensitivity results are slightly different. It is also observable that with increasing size of parameter perturbation, the number of parameter sets which need to be sampled are increasing. Therefore, small parameter perturbations should preferably be applied to minimize the calculation costs since larger perturbations yield effectively no gain in information. If sticking to small perturbations, the results of the sensitivity analysis can be considered to be independent from the actual value and direction of the parameter perturbation.

2.5.5. Statistical evaluation and data interpretation

In the sensitivity analysis of each model, it is sampled until for 2500 parameter sets the period and amplitude sensitivity has been determined. The resulting period and amplitude sensitivity distributions are further characterized with the help of descriptive and comparative statistics. In the statistical evaluation, this work relies on measures based on rankings of data rather than their particular values. The median is used instead of the arithmetic mean value as measure of central tendency. As already mentioned in section 2.4.1, this choice has been made due to the non-normal, skewed and heavy-tailed distributions of the sensitivity data obtained in the analyses. Why medians are preferably used is illustrated additionally in the Appendix where it is shown that the arithmetic mean is highly sensitive to extreme values and does not reflect the actual central tendency of the data (section A.1.5). The characteristics of the sensitivity distributions are captured in special box-plots (section 2.4.3). Examining more than one parameter set for each model enables assessing the potential of the models as to which period and amplitude sensitivity values and combinations are possible. The correlation between amplitude and period sensitivity is determined with Spearman's rank correlation coefficient R_S (section 2.4.4). The significance of differences between sensitivity distributions of different models is evaluated with the help of the MWU test (section 2.4.5) which as being a non-parametric test is considered most appropriate here due to the data not following a particular distribution.

3. The influence of the structure on robustness

In this chapter, it is studied whether the structure of the model of a biological process plays a role for the period and amplitude sensitivity. The alternative would be that it is only the choice of the parameter set for the model that influences the obtained period and amplitude sensitivities. For this purpose, calcium and circadian oscillations are analyzed which are two biologically well-examined oscillating systems (section 3.1). In their biological context, the two oscillatory processes differ critically in their period sensitivity properties. For each of the two biological processes, an exemplary model is chosen and a sensitivity analysis is performed (section 3.2). Based on the results for calcium and circadian oscillations, different features of the model structure are listed and described (section 3.3). Their effects on period and amplitude sensitivity are examined in the following chapters.

3.1. Circadian and calcium oscillations

3.1.1. Circadian oscillations

Circadian oscillations are ubiquitous in biology, occurring in bacteria, plants and a plethora of, but not all, animals (see reviews in Johnson et al. [2008], Leloup [2009], Jolma et al. [2010], Hogenesch and Herzog [2011], Goldbeter et al. [2012]). The name comes from the Latin *circa* - around and *dies* - day indicating that the period of the oscillations is about 24h. The oscillations ensure reliable time-keeping in organisms enabling the anticipation of the daily night-day rhythm induced by earth rotation.

Due to their high prevalence, circadian oscillations are subject to intense experimental and theoretical studies and have been characterized well. Typical organisms whose circadian rhythms are studied in great detail are mammals, the fruit fly (*Drosophila*), plants (especially mouse-ear cress, *Arabidopsis*) and fungi (e.g. *Neurospora*), but also cyanobacteria.

In general, transcriptional gene regulation is acknowledged to play a central role in the generation of circadian oscillations (Wagner [2005], Gerard et al. [2009], Leloup [2009], Jolma et al. [2010], Goldbeter et al. [2012], Teng et al. [2013]). In all model organisms listed above, a negative transcription-translation feedback loop has been found on cellular level which is supposed to be the basis for the circadian oscillations. This means that a particular "clock protein" inhibits the transcription of its own gene (see the scheme in Figure 3.1 A). The resulting negative feedback loop encompasses at least the mRNA of this clock protein which is produced by transcription from the gene and the clock protein itself which is produced from the mRNA by translation, and some mediating protein which is inhibited by the clock protein and acts as transcription factor activating the transcription of the clock protein gene. Higher organisms often yield multiple clock proteins. The clock proteins being at the core of this feedback loop are for example isoforms of period circadian protein homolog (per) and cryptochrome (cry) in mammals (Becker-Weimann et al. [2004], Baggs et al. [2009]), period (per) and timeless (tim) in *Drosophila melanogaster* (Goldbeter [1995], Leloup [2009]), FRQ in *Neurospora crassa* (Cheng et al. [2001], Jolma et al. [2010]) and KaiC in cyanobacteria (Kitayama et al. [2008], Teng et al. [2013]). For *Arabidopsis*, it is an interplay of at least three proteins acting on the transcription of the others which form the core negative transcription-translation feedback loop (Locke et al. [2005a], Jolma et al. [2010]).

Additional regulatory patterns and complementary mechanisms for generating oscillations are known to exist, but they differ considerably between organisms. In mammals, for example, a second feedback loop via rev-erbα is installed (Becker-Weimann et al. [2004], Baggs et al. [2009]); in cyanobacteria, a biochemical oscillation system composed of the proteins KaiA, KaiB and KaiC is present (Johnson et al. [2008], Teng et al. [2013]).

Circadian oscillations persist in general at single cell level. For higher organisms, coupling of oscillations between cells and special "zeitgeber" tissues as the suprachiasmatic nucleus (SCN) in mammals have been identified (Baggs et al. [2009], Jolma et al. [2010], Hogenesch and Herzog [2011]). Although coupling effects may be im-

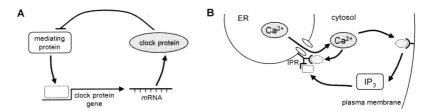

Figure 3.1.: Schemes of the central feedback loops of circadian and calcium oscillations. A: In circadian oscillations, a negative transcription-translation feedback loop is established by a clock protein which is produced by translation from its mRNA. The protein inhibits some mediating protein which is an activating transcription factor for the transcription of the clock protein. Figure adapted from Dunlap [1998]. B: The core feedback of calcium oscillations is a calcium induced calcium release. Calcium from the cytosol activates production of IP$_3$ at the plasma membrane. IP$_3$ binds to a receptor (IPR) at the ER membrane thereby triggering the release of calcium from the ER to the cytosol. Cytosolic calcium directly increases the flow of calcium from the ER to the cytosol through the calcium channel coupled to the IPR.

portant for the characteristics of the circadian oscillations in these organisms (Gonze et al. [2005], Locke et al. [2008], Bordyugov et al. [2013]), this work restricts the analysis to circadian oscillation models of single cells containing a negative transcription-translation feedback loop which can rely on a long history of scientific study and which is considered to exist in many organisms.

3.1.2. Calcium oscillations

Calcium plays a crucial role in a wide variety of cells and tissues in mammals, for example in muscle and immune cells, neurons, the liver, and oocytes, and participates in a multiplicity of physiological functions (reviewed e.g. in Berridge et al. [2000], Dupont et al. [2011], Bootman [2012], Goldbeter et al. [2012]). Calcium acts as second messenger, modulates the activity of cell motility, numerous enzymes as kinases and phosphatases and transcription factors as for example NF-κB (Berridge et al. [2000]). Alterations in cytoplasmic calcium can mediate extremely rapid events as muscle contraction or exocytosis, but also slower responses as cell division, differentiation, and apoptosis (Dupont et al. [2011]). Periodic changes in calcium concentrations can arise spontaneously or by electrical, mechanical or chemical stimulation with hormones or neurotransmitters (Goldbeter [2002], Bootman [2012]). Thereby, calcium enters the cytoplasm through channels from the extracellular medium and from intracellular calcium stores as the endoplasmic (or sarcoplasmic) reticulum (ER, Berridge et al. [2000], Sneyd et al. [2004], Dupont et al. [2011], Bootman [2012]).

This work focuses on models of slower periodic alterations in cytosolic calcium concentration, here referred to as calcium oscillations. They arise in not electrically excitable cells and their period ranges from tens of seconds to several minutes (Dupont et al. [2011], Goldbeter et al. [2012]). In contrast, rapid events mediated by calcium are regulated in electrically excitable cells which display regenerative all-or-none plasma membrane channel activations, so-called calcium action potentials (Dupont et al. [2011]). Such rapid calcium transients lasting only hundreds of milliseconds occur for example in cardiac myocytes (Bootman [2012]). Their examination is not the focus of this thesis.

Calcium oscillations rely on a sequential regenerative discharge of stored calcium (Dupont et al. [2011], Bootman [2012]). The most prominent mediating protein for release of calcium from the ER is inositol 1,4,5-triphosphate (IP$_3$, Berridge et al. [2000], Sneyd et al. [2004], Dupont et al. [2011], Goldbeter et al. [2012], scheme in Figure 3.1 B). Calcium binds to plasma membrane receptors mediating the synthesis of IP$_3$. IP$_3$ then diffuses through the cytoplasm, binds to specific receptors (IP$_3$ receptors, IPR) and triggers the release of calcium from the ER (Sneyd et al. [2004]). The resulting increase in cytosolic calcium is amplified by an intracellular calcium release through calcium-sensitive calcium channels in the store (ER) membrane, a positive feedback known as calcium-induced calcium release (Berridge et al. [2000], Haberichter et al. [2001], Goldbeter [2002], Sneyd et al. [2004], Dupont et al. [2011]). The principle of calcium-induced calcium release forming a positive feedback is present even twice here: Indirectly by calcium activating IP$_3$, and directly as calcium itself increases the open probability of the calcium channels coupled to the IPR. ATP-dependent pumps (sarcoplasmic or endoplasmic reticulum ATPases) are antagonizing the positive feedback returning calcium from the cytosol back to the intracellular stores (Camacho and Lechleiter [1993], Berridge et al. [2000], Goldbeter [2002], Bootman [2012], not depicted in the scheme in Figure 3.1 B).

3.1.3. Differences between circadian and calcium oscillations

Period robustness properties

The oscillatory characteristics of calcium and circadian oscillations differ. The circadian rhythm displays a very robust period in experiments. The oscillation period remains nearly unaffected by changes in environmental conditions as temperature, pH, and nutritional conditions (Pittendrigh [1954], Pittendrigh and Caldarola [1973], Ruby et al. [1999], Ruoff et al. [2000]). The underlying reason for the biologically observable robustness of the period, also towards other types of variations than temperature, is not yet fully understood (Jolma et al. [2010], Hogenesch and Ueda [2011]). Hypotheses for temperature compensation have been proposed. One hypothesis assumes that each of the enzyme-driven reactions is temperature compensated itself (Hogenesch and Ueda [2011]). Another hypothesis is the balancing reactions theory proposing that perturbations of the period are canceled by opposing influence of altered parameter values on reaction velocities and hence the period. This is observed in experiments and may be the case for *Neurospora crassa* (Hogenesch and Ueda [2011]). However, this mechanism would need fine-tuning of the kinetic parameters (Ruoff et al. [2000, 2003]).

In contrast to circadian oscillations whose periods remain widely unchanged for alterations of environmental conditions, the period of calcium oscillations strongly varies with external conditions. It is highly responsive to changes in temperature and agonist concentration (Rooney et al. [1989], Schipke et al. [2008]). Depending on the cell type, the period reaches from seconds to minutes (Dupont et al. [2011], Goldbeter et al. [2012]). While circadian oscillations are considered to provide a time-keeping function by its period, the function of calcium oscillations is discussed to lie in the encoding of signals in terms of period or frequency. A variety of physiological responses are controlled by the frequency (and waveform) of calcium oscillations, e.g. smooth airway muscle contractions (Sanderson et al. [2008]), kinase activity (De Koninck and Schulman [1998]) and gene expression profiles (Dolmetsch et al. [1998], Cai et al. [2008], Paszek et al. [2010]). In summary, circadian and calcium oscillations display two extreme cases of period robustness properties, the first having a very robust, the second a very sensitive period.

Structural properties

In addition to the differences in the biologically observed period sensitivities, also the structures of the biological processes underlying circadian and calcium oscillations differ. First, there are different basic types of feedback structures. Negative feedback is considered to define the circadian clock oscillations, generally by a transcription-translation feedback loop including the genes participating in the process (Aronson et al. [1994], Goldbeter [1995], Wagner [2005], Baggs et al. [2009], Teng et al. [2013]). In contrast, calcium oscillations generally rely on a positive feedback that includes the enhancement of the flow of endoplasmic calcium into the cytosol induced by cytosolic calcium (Goldbeter [2002], Sneyd and Falcke [2005], Dupont et al. [2011]). Second, the processes differ in the structure of the conversion of matter. In calcium oscillations, reactions generally represent transport processes where all matter is converted from one species to another. This is not the case in circadian oscillations. The basic transcription-translation feedback loop of circadian oscillations incorporates the expression of at least one gene including the transcription and translation steps. In the process of protein production, the the DNA is not converted to the mRNA, and the mRNA is not converted to the protein, but both only influence the respective production rate. Therefore, in circadian oscillations, the matter flow between species is disrupted.

3.2. Model structure or choice of the parameter set

The structural properties of an examined biological process are in general the basis for the curation of the models of the process and hence the model structures represent the structural properties of the process. Additionally, the period robustness differences in circadian and calcium oscillations described in section 3.1.3 shall be reflected by the models of the process in question. In Wolf et al. [2005], the period sensitivity of several circadian and calcium oscillations models has been examined at the reference parameter set, which is the parameter set found to reproduce the experimental observation which is published together with the model. Therein, the periods of oscillations of models describing circadian rhythms have been found to be more robust than the period of models describing calcium oscillations. Without further examination, however, it is not immediately clear whether these differences in period sensitivity arise from differences in the model structures comparing the examined calcium and the circadian oscillations models or from the choice of the particular reference set of parameter values used for the estimation of the period sensitivity.

If the contribution of the choice of the parameter set on period and amplitude robustness was large compared to the effect of the structure of the model, it is neither feasible nor of interest to identify structural properties affecting period robustness as it is the aim of this work. *A priori*, it is not clear that the model structure has to influence a certain property of a system. For example, with the help of mathematical modeling it was shown that for a particular structural motif occurring in yeast transcriptional regulation networks, the resulting gene expression pattern depends only on the choice of the particular parameter set (Ingram et al. [2006]). A characteristic output for the motif itself could not be identified. On the contrary, it has been described that in an artificial oscillator model structural properties as the type of feedback employed play a role for the oscillatory properties and their variability (Tsai et al. [2008]).

Therefore, it is now assessed to which extent the following two sources are responsible for the period robustness differences in circadian and calcium oscillation models: The underlying structure of the model as feedbacks, matter flow or kinetics in some of which the processes are known to differ (section 3.1.3) and/or the particular choice of the values of the kinetic parameters as enzymatic binding rates and transport rates. For this purpose, sensitivity analyses as described in section 2.3 are performed. They rely on random sampling and lead to calculation of the period sensitivity values σ_T and amplitude sensitivity values σ_A (as described in section 2.2) for 2500 distinct parameter sets for each examined model. The obtained sensitivity distributions of 2500 data points give an impression of the potential of the models regarding period and amplitude robustness. It can be assumed that the model structure plays a role for sensitivity if models yielding different structural properties display different sensitivity distributions. Differences can be assessed by comparing central tendencies of the sensitivity distributions as well as their variation, and correlation between period and amplitude sensitivity (compare methods section 2.4).

3.2.1. The chosen models

For circadian and calcium oscillations, several mathematical models have been proposed (reviews in Leloup [2009], Sneyd and Falcke [2005], more models in the BioModels Database, Le Novere et al. [2006], Li et al. [2010]). They differ in the choice of rate laws, number of variables, and feedback interactions. For this analysis, one basic model is examined for each process capturing structural key properties as the negative transcription-translation feedback loop and additional regulation for the circadian rhythms and the calcium-induced calcium release for calcium oscillations. A mammalian circadian oscillation model proposed by Becker-Weimann et al. [2004] and a calcium oscillation model developed by Goldbeter et al. [1990] are chosen. The ODE systems of the models are given in the Appendix, section A.7.1.

The mammalian circadian oscillations model (Becker-Weimann et al. [2004]) is composed of a negative transcription-translation feedback loop consisting of per/cry mRNA and the cytosolic and nuclear proteins. An additional regulatory feedback loop includes the positive effect of brain and muscle aryl hydrocarbon receptor nuclear translocator-like 1 (BMAL1) which is indirectly established via rev-erbα. In the model, BMAL1 mRNA is included as well as the cytosolic and nuclear protein concentrations. Activated nuclear BMAL1 positively influences per/cry mRNA production. All in all, the model comprises seven species, connected by 17 reactions, including four nl-parameters and three cooperativity parameters.

The calcium oscillations model (Goldbeter et al. [1990]) consists of only two species, cytosolic and endoplasmic calcium. It is a phenomenological model for signal-induced calcium oscillations. The important regulator IP_3 for the calcium induced calcium release contributes as a parameter reflecting the degree of the external stimulus signal strength. The model considers two calcium stores, one which is IP_3-sensitive and another one being sensitive to calcium. It comprises two species, six reactions, three nl-parameters and three cooperativity parameters.

3.2.2. Results of the sensitivity analysis

Both models describing circadian and calcium oscillations are analyzed with respect to their period and amplitude sensitivity properties according to the method described in section 2.3. Figure 3.2 shows the amplitude sensitivity values versus the period sensitivity values. The red triangles display the results for the circadian rhythm model, the blue circles those for the calcium oscillations model. Each of the 2500 dots of each color represents the sensitivity data for one specific sampled parameter set for the respective model. The black symbols denote the median values. The inset displays the sensitivities for parameter sets of both models yielding sensitivity values larger than 100. The characteristics of the sensitivity distributions as the median, the 95% CI of the median, the range between first and third quartile, and 90%R are graphically captured in the box-plots (section 2.4.3) on top (period sensitivity) and to the right (amplitude sensitivity) of the scatter plot. The numerical values of the distribution characteristics are given in Table 3.1 as well as the Spearman's rank correlation coefficient R_S (section 2.4.4) and the total number of sampled parameter sets (# sets) in order to obtain 2500 parameter sets for which

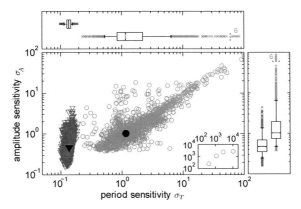

Figure 3.2.: Sensitivity analysis of a circadian and a calcium oscillations model. Period and amplitude sensitivities are computed from Equations 2.6 with a perturbation of 2% of each parameter for a calcium model (blue circles, Goldbeter et al. [1990]), and a circadian model (red triangles, Becker-Weimann et al. [2004]). Each point gives the sensitivities for a different parameter set (section 2.3), black symbols denote the median values. The inset depicts the results for parameter sets with sensitivity values larger than 100. The box-plots show the characteristics of the sensitivity distributions for period sensitivity (top) and amplitude sensitivity (right) for the circadian (red) and calcium model (blue).

Table 3.1.: Table for the sensitivity values of a circadian rhythm model (Becker-Weimann et al. [2004], id "circ") and a calcium oscillations model (Goldbeter et al. [1990], id "ca"). Given are the period (per.) and amplitude (ampl.) sensitivity distribution medians (μ), the 95% confidence interval of the median (95% CI), the 25th, 75th, fifth, and 95th percentiles (Q_p, forming their respective intervals), the 90%R, Spearman's rank correlation coefficient when comparing the period and amplitude sensitivities of the model (R_S), and the number of sampled parameter sets in order to find 2500 oscillating sets for the analysis (# sets). Data corresponds to Figure 3.2.

id	sens	μ	95% CI	$[Q_{25},Q_{75}]$	$[Q_5,Q_{95}]$	90%R	R_S	# sets
circ	per.	0.14	[0.14,0.14]	[0.13,0.15]	[0.11,0.16]	0.05	0.52	$6.1 \cdot 10^5$
	ampl.	0.47	[0.46,0.48]	[0.35,0.68]	[0.24,1.37]	1.13		
ca	per.	1.16	[1.14,1.17]	[0.85,2.22]	[0.53,6.45]	5.92	0.82	$7.1 \cdot 10^4$
	ampl.	1.01	[1.00,1.03]	[0.74,1.94]	[0.46,6.54]	6.08		

the sensitivity has been calculated.

From Figure 3.2, it can be observed that the paired sensitivity distributions of the circadian and calcium oscillation model are not overlapping. This is mainly caused by the period of the oscillations in the calcium signaling model (blue) being systematically more sensitive to parameter perturbations than in the circadian rhythm model (red, see Figure 3.2 scatter plot and box-plot on top). The median period sensitivities of the two models are decisively different, the numerical values being 0.14 for the circadian and 1.16 for the calcium oscillation model, respectively. Even the 90% data ranges do not overlap (Table 3.1). The detailed results of the MWU test (section 2.4.5) comparing the period sensitivity values of the circadian and the calcium oscillation model are given in Table 3.2. The p-value obtained for the period sensitivity is zero, which is the lowest value possible. Hence, the null hypothesis of circadian oscillations and calcium oscillations having the same period sensitivity distribution can be rejected with arbitrarily high confidence. The difference is highly significant. From this, one can conclude that the structures of the models may shape the period sensitivity properties of the models. The model structure of the circadian model thereby allows for rather low period sensitivities whereas the calcium model structure yields rather high period sensitivity values.

It is now focused on the impact of the choice of the parameter set on period sensitivity displayed by the width of the period sensitivity distribution. The circadian oscillations model (red) reveals a smaller range than the calcium oscillation model

Table 3.2.: Results for the MWU test comparing the circadian (Becker-Weimann et al. [2004], id "circ") and the calcium oscillations model (Goldbeter et al. [1990], id "ca"). Period (per.) or amplitude (ampl.) sensitivity (sens.) distributions are compared. The MWU p-value (MWU p), the numbers of parameter sets compared (n_X, n_Y), the U- and the z-value (z-val) are given. Data corresponds to Figure 3.2.

id$_X$		id$_Y$	sens.	MWU p	n_X	n_Y	U	z-val
circ	$\Leftrightarrow$	ca	per.	0	2500	2500	3126250	-61.23
circ	$\Leftrightarrow$	ca	ampl.	0	2500	2500	4105063	-42.05

(blue). Indeed, the 90%R span 0.05 and 5.92 for the circadian and the calcium oscillations model, respectively (Table 3.1). This means that the interval where 90% of the period sensitivity data are located 90%R is nearly 120-fold higher for the calcium model than for the circadian model. This leads to the conclusion that the period sensitivity of the calcium oscillations model can be by far more easily tuned by the choice of the parameter set than this is the case for the circadian oscillations model.

For the sensitivity values of the amplitude, the differences between the two models are quite similar as for the period sensitivities but less pronounced. As exhibited in the box-plot to the right in Figure 3.2, the ranges of the amplitude sensitivity values overlap slightly, even though they do not at the level of the first and third quartiles. The median amplitude sensitivity of the circadian model is less than half the median amplitude sensitivity of the calcium model (0.47 and 1.01, respectively, Table 3.1), the MWU test gives again a p-value of zero (Table 3.2). This leads to the conclusion that the model structure also has a strong impact on the amplitude sensitivities of the models; the circadian model structure gives rise to rather low, the calcium model structure to rather high amplitude sensitivity values.

To compare the widths of the amplitude sensitivity distributions, the 90%R is considered. It spans 1.13 and 6.08 for the circadian and the calcium model, respectively (Table 3.1). Thus, the 90%R of the amplitude sensitivity distribution of the calcium model is 5.4-fold than that of the circadian model. This indicates that not only the period sensitivity but also the amplitude sensitivity of the calcium oscillations model can be more easily tuned through the choice of the parameter set than that of the circadian oscillations model. Thereby, the circadian model reaches to lower sensitivity values.

More information captured by the sensitivity analysis refer to the relation between the period sensitivity and amplitude sensitivity in each model. In the circadian model, the amplitude can be more easily tuned via parameter choice than the period sensitivity (90%R: 0.04 and 1.12, respectively). For the calcium model, the period and the amplitude sensitivity can be similarly tuned via the choice of the parameter set (90%R: 5.92 and 6.08, respectively). The results of the distribution widths partially accord with the rank correlation coefficients R_S (values see Table 3.1) indicating how strongly the period and amplitude sensitivities are correlated. It reveals an at most moderate positive correlation between period sensitivity and amplitude sensitivity in the circadian model ($R_S = 0.52$) which means that by the choice of the parameter set, the amplitude sensitivity can be tuned rather independently from the period sensitivity. This may be also due to the very narrow distribution range for the period sensitivity. In contrast, in the calcium model, a strong positive correlation between period and amplitude sensitivity is observed ($R_S = 0.82$). This means that if increasing the period sensitivity by choosing another parameter set, it is highly probable that also the amplitude sensitivity is increased and *vice versa*.

3.3. Summary and discussion: The influence of the structure on robustness

This chapter aims to assess whether structural properties of a model of a biological process considerably affect its period sensitivity and amplitude sensitivity and hence whether they need to be considered for sensitivity predictions. For this purpose, the comparison of calcium and circadian oscillations proves useful for three reasons: First, both systems have been extensively studied regarding their structure, and one can rely on a variety of established models. Second, the two processes are found to differ in structural properties as feedback type and matter flow properties (section 3.1.3). Third, the impact of structural properties might be more easily detected since the two examined processes display extreme biologically observed differences in period robustness (section 3.1.3).

Therefore, one model for circadian (Becker-Weimann et al. [2004]) and one model for calcium oscillations (Goldbeter et al. [1990]) are examined which capture the key structural characteristics of the underlying process (section 3.2.1). A sensitivity analyses reveals that the sensitivity distribution clouds of the two models do not overlap (section 3.2.2). This envisions that the period and amplitude sensitivities of the chosen calcium model as well as the circadian model are strongly influenced by the structural properties of the models. Thereby, the model structure of the calcium model leads to rather high period sensitivity values, the model structure of the circadian model allows only for low period sensitivity values. These tendencies have also been observed in the examination of the period sensitivities of several calcium and circadian oscillation models for one particular parameter set (Wolf et al. [2005]). The time-keeping function of circadian oscillations requires a stable period which is reflected by a highly conserved low period sensitivity value for any choice of one of the parameter sets sampled here. In contrast, calcium oscillations are discussed to act in frequency-encoded signaling, therefore the period needs to be variable and the period sensitivity should be high as is observed for the majority of parameter sets in the sensitivity analysis. One can conclude that the majority of the sampled parameter sets in each model allows for the necessary period sensitivity to reflect its supposed function of the oscillation period of its process.

In both models, the amplitude sensitivity can take low and also rather high values. In fact, it is known that in contrast to the period, the amplitudes of the components in the circadian oscillations are variable if exposed to environmental changes as temperature (Jolma et al. [2010]). Therefore, increased amplitude sensitivity may be expected for the circadian oscillation model. For calcium oscillations, there are indications that the amplitude of the calcium spikes are rather fixed in many cell types, and the mitochondria have been found to be decisive for this amplitude robustness (Grubelnik et al. [2001]). In the examined example model, the amplitude is not robust to perturbations, which might be due to the use of a minimal model for the examination which was not designed to display the feature of a constant amplitude for the arising oscillations.

Interestingly, especially for the period, the overall results obtained in the sensitivity analyses also persist if altering the sensitivity measure by neglecting the effect of nl-parameters and summing only over the sensitivity coefficients of the rate coefficients or by summing only over the three largest sensitivity coefficients (Appendix, section A.1.2). This strengthens the conclusion that at least the period robustness is strongly affected by the general structural principles of the system as for example those described in section 3.3.1 whose influence on period and amplitude sensitivity is examined more thoroughly in this work.

The results of the sensitivity analyses for the two models fit the biological observations of circadian and calcium oscillations very well. However, this might be caused by the choice of the models. Whether general conclusions may be drawn is further researched in section 7.2 where two more models of each process are investigated.

3.3.1. Structural properties to be examined in this work

In section 3.2, it is shown that the structure of a system strongly affects the (period) sensitivity properties. Based on the structural differences of the biological processes underlying circadian and calcium oscillations (section 3.1.3), in the following, three properties which might affect period and amplitude robustness are described. The effects of these three properties on period and amplitude sensitivity are examined in the course of this work with the help of a chain model of length four as prototype oscillator (Wolf et al. [2005], section 4.2).

(i) The effect of a negative or positive feedback on period and amplitude sensitivity are studied (chapter 4). It has been discussed that the type of feedback might impact robustness (Tsai et al. [2008], Stricker et al. [2008]), and calcium and circadian oscillations differ in their basic feedback structure.

(ii) The consequences of the occurrence of reactions with or without flow of matter between species are investigated (chapter 5). Calcium oscillations generally rely on reactions including matter flow between species whereas matter flow is disrupted in circadian oscillations by at least the reaction encoding the transcription or translation.

(iii) It may be important for the robustness of period and amplitude whether reactions can saturate, as for example reactions with Michaelis-Menten kinetics, or whether they follow linear mass action kinetics. With respect to this feature, there is no favor for the occurrence of the one or the other kinetics in calcium or circadian oscillation models. Nevertheless, the effects of using saturating instead of linear kinetics on the sensitivity of oscillatory properties remain widely unexplored in literature and are treated in this work (chapter 6).

4. The effect of feedbacks on sensitivity

In this chapter, first, feedbacks and their biological function are described in general (section 4.1). Then, the structure of the chain models of length four is introduced (section 4.2), it is presented how the feedbacks are implemented in the chain models (section 4.2.1), and a stability analysis is performed (section 4.2.2). The results of the sensitivity analysis of the chain models with mass action kinetics and a positive or a negative feedback are compared (section 4.3). An analysis of maximal fluxes in the chain model with negative feedback and bifurcation analyses for special parameter sets in both models provide hints on the origins of low or increased sensitivities (section 4.4). Furthermore, they illustrate the occurrence of birhythmicity for the positive feedback chain model.

4.1. Feedbacks and feedback loops in general

4.1.1. Feedbacks and feedback loops in biology

Feedbacks are ubiquitous in engineering as probably in biology since they deliver huge benefits in a variety of processes (Csete and Doyle [2002]). Feedbacks are indispensable for regulation or noise-reduction, i.e. the reduction of perturbations caused by changes in molecule abundances. Additionally, they decisively enlarge the variety and complexity of possible system behavior, e.g. by enabling bifurcations to occur (Nguyen and Kulasiri [2009]). Biologically, a feedback denotes an effect from a species concentration known to be further down in a reaction chain (downstream) to a species concentration or reaction being more up in a reaction scheme (upstream). The feedback is considered to be positive if the effect of the downstream species on the upstream species is positive, and negative if the effect is negative. For example, a negative feedback can be established by the downstream species decreasing the production or by increasing the degradation of the upstream species.

A feedback establishes a loop of interactions between species: a feedback loop. This means that changes in the concentration of each species in this loop indirectly further influence its own concentration, the species feed back on themselves. A feedback loop can be either positive, negative, or ambivalent, i.e. switch between positive and negative upon concentration or parameter changes. Both positive and negative feedback loops have been subject to various studies. For positive feedback loops, the occurrence of bistability is a well-known feature (Monod and Jacob [1961], Griffith [1968b], Ferrell [2002], Novak and Tyson [2008]). Besides the creation of bistable switches, two further common functions in signaling of a single positive feedback loop are (i) amplifying a signal or (ii) changing the timing of a signaling response (Brandman and Meyer [2008]). Negative feedback loops are essential for regulation, reduce noise and hence stabilize the steady state (Stelling et al. [2004a]), but thy can also amplify time-varying perturbations (Csete and Doyle [2002]). For a single negative feedback loop, one can distinguish four signaling functions: (i) stabilizing steady states, (ii) limiting the output, and establishing either an (iii) adaptive or a (iv) transient response (Brandman and Meyer [2008]). The characteristics and initial conditions of the system under investigation determine which of the functions is realized. Sustained oscillations which are the focus of this work can only occur if the system possesses a negative feedback loop which drives the system back to its state of the beginning (e.g. Griffith [1968a], Novak and Tyson [2008]). Additionally, a certain delay in the negative feedback loop is required for oscillations (Stricker et al. [2008], Hogenesch and Ueda [2011]). The delay can be increased by longer reaction chains through multi-step processing (Novak and Tyson [2008]), saturated degradation or switch-like responses where a species must reach a threshold concentration (Mengel et al. [2010]) which can be implemented by stronger cooperativity (Wagner [2005]).

4.1.2. Known effects of feedbacks on oscillatory properties

Different effects of feedback loops on oscillatory characteristics have been investigated using different methods. Since a negative feedback loop is the requirement for sustained oscillations, primarily the effects of additional feedback loops can be estimated. For example, in a three-species model with negative feedback, the amplitude of the oscillations has been shown to be larger if additionally introducing a positive feedback (Tian et al. [2009]). Positive feedback in phosphorylation and dephosphorylation reactions has been shown to enlarge the amplitude of cell-cycle oscillations as well as the oscillation probability (Gerard et al. [2012]).

Advances to detect the effects of different feedback types on the sensitivities of period and amplitude have been made. In Zamora-Sillero et al. [2011], a model with a positive feedback, establishing a negative and a positive feedback loop, and a negative feedback, establishing only a negative feedback loop, has been examined. Although it has been distinguished between "essential" and "non-essential" feedbacks, both types of feedback are present together throughout the published analysis, and it is infeasible to distinguish between effects caused by each individual feedback. Additionally, the robustness which has been examined is very different from the definition of sensitivity as employed in this thesis. The notion of "viable region" in parameter space is employed, referring to oscillations with a period in a certain interval (Zamora-Sillero et al. [2011]). As robustness measures, the size of the viable region and the lengths of random walks until the viable region is left are used. For both measures, it has been concluded that essential negative feedback models are most robust concerning fixed-period oscillations. However, it has been also observed that more parameter sets are viable the stronger the positive feedback, and it has been concluded that a positive feedback increases this type of robustness of the essential negative feedback architecture (Zamora-Sillero et al. [2011]).

Also the investigations in Tsai et al. [2008], Stricker et al. [2008], Nguyen [2012] on the effect of feedbacks on oscillatory properties do not employ sensitivity coefficients, but the concept of "tunability" as indicator of the size of the operational range of the period or amplitude. The operational range is defined as the interval in which the period or amplitude (of a particular species) varies for variations in a parameter (a particularly chosen parameter in Tsai et al. [2008], an activator concentration in Stricker et al. [2008], the feedback strength in Nguyen [2012]) over the whole range of oscillatory behavior. However, "tunability" as in the above mentioned publications and "sensitivity" as employed in this work are not necessarily equal. If, for example, the period of a model for a particular parameter set is very tunable, this does not automatically mean that it reacts more strongly to the same parameter perturbation than for a parameter set exhibiting a less tunable period. Additionally, the sensitivity employed in this work considers variations with respect to *all* kinetic parameters of the models instead of to only few or one selected. Thus, the two notions "sensitivity" as employed in this thesis and "tunability" as employed in Tsai et al. [2008], Stricker et al. [2008], Nguyen [2012], are two facets of the robustness of oscillatory properties that provide different insights.

In three-species models, the effect of combinations of two negative feedback loops (the outer-loop of length three and a shorter inner-loop of length two or one) has been investigated (Nguyen [2012]). The author has found that the period can be tuned by the feedback strength, the period increases with stronger outer-loop feedback strength and decreases with increasing inner-loop feedback strength. Furthermore, a decoupling of function has been identified: altering the inner-loop feedback strength modulates the period or amplitude while keeping the mean value (of the third species) constant, altering the outer-loop feedback strength can modulate the period while keeping the amplitude (of the third species) almost constant.

Also the tunability of the period and amplitude in models with only negative to negative-plus-positive feedback loops have been compared (Tsai et al. [2008], Stricker et al. [2008]). In Tsai et al. [2008], a three-node oscillator based on conversion cycles has been examined. The model yielding only a negative feedback loop exhibits only minor changes in period while having a variable amplitude. For the same model with an additional autoregulatory positive feedback loop, a tunable period and constant amplitude for changing the one particular parameter has been found. Their investigation of the basic oscillator model includes a parameter sampling procedure, thus the results are valid for multiple parameter sets. The same pattern of tunable amplitude and stable period for various negative-feedback-only models, and tunable period for fixed amplitude for various negative-plus-positive-feedback models has been observed at one parameter set for each model (Tsai et al. [2008]). Furthermore, in the three-node basic oscillator models considered in Tsai et al. [2008], positive feedback has been found to enhance the oscillation probability.

Similar results concerning the period tunability have been derived for a synthetic oscillator implemented in *Escherichia coli* and the according stochastic and deterministic models for a single parameter set only (Stricker et al. [2008]). The oscillator with a positive feedback loop allows for greater variability of the period than the oscillator with a single negative feedback loop. The period is tunable by e.g. levels of a certain activator, temperature or media source.

In the presented thesis, a large number of parameter sets is sampled and changes in all kinetic parameters are considered for the calculation of the sensitivities of the period and amplitude. This section deals with the general effects of negative or positive feedback on the overall sensitivity of the period and amplitude of oscillatory processes. The relation between the herein obtained and the published results is discussed in section 4.5.1.

4.1.3. Feedback vs. feedback loop in ODE systems

The biological description of feedbacks and feedback loops given in section 4.1.1 is rather vague. A more formal definition of feedback loops is possible for ODE models (Thomas and Kaufman [2002], section 2.1.4). Mathematically, a feedback loop, also referred to as feedback circuit, is considered as a closed set of oriented interactions between species. Only the shortest possible loops are considered, i.e. each species may occur only once on each side of the interaction. All feedback loops of an ODE model can be found by inspecting its Jacobian matrix (see section 2.1.4).

Frequently, the notions of feedbacks and feedback loops are used synonymously in biology, but their relation is in general not one to one. First of all, a negative or positive feedback may result in a positive or negative feedback loop, respectively. This is for example the case if the number of negative interactions in the interaction chain between the upstream and the downstream species is uneven. Second, a feedback may establish more than one feedback loop. In ODE models, feedbacks are encoded as species acting on particular reactions (see section 2.1.4). Thus, one feedback can result in more than one interaction if the reaction which is regulated by the feedback affects multiple species. This is the case if the reaction possesses e.g. one product and one source species, or multiple product or source species. Indeed, one feedback generates exactly one feedback loop if two requirements are satisfied: (i) only one species is affected by the reaction being influenced by the feedback and (ii) only one directed chain of interactions between the regulated and the regulating species exists. Requirement (i) is exactly met if the reaction influenced by the feedback has only one either source or product species. This can be directly read off from the stoichiometric matrix which then includes only exactly one non-zero entry for the according row of the reaction.

By inspecting the Jacobian matrix, also feedback loops are revealed which would normally not be considered as feedback loops. For example, reversible transport reactions from one compartment to another are always establishing positive feedback circuits. However, in biology, a back-transport reaction without obvious regulatory actions on it would be rarely considered as a positive feedback. Even requesting some sort of nonlinearity in the feedback loop found by the Jacobian matrix in order to make it a valid, biologically accepted feedback would not solve the problem. Indeed, as soon as not assuming a mass action kinetics transport, but e.g. a Michaelis-Menten rate for the back-transport, there is a non-linearity, but still the transport back and forth would rather not be considered as feedback (loop) in a biological process.

Therefore, the feedback loops found in an ODE model by examining the Jacobian matrix do not necessarily coincide with the feedbacks or feedback loops considered in biology. Hence, although widely employed, the terms "feedback" and even "feedback loop" are not uniquely defined in biology. In this work, the term "feedback" refers to the general, biological interpretation as described in the beginning of this section. The structural or mathematical implications on the (type of) feedback loops or feedback circuits as obtained by analyzing the Jacobian matrix are explained and considered where appropriate.

4.2. The chain models

In order to examine the three structural principles described in section 3.3.1, feedback type, matter flow properties and reaction kinetics, the core models of a chain of length four is used (Wolf et al. [2005]). They are composed of a chain of four species, $S_1, \ldots, S_4$, where the fourth is feeding back on the reaction from the first to the second, ν_2, either positively or negatively (Figure 4.1, ODE systems in the Appendix, section A.7.2). Each of the four species is subject to degradation. Different chain lengths are discussed in section 4.5.2. The model is referred to as negative feedback chain model or positive feedback chain model, respectively, if the action of S_4 on ν_2 is negative, i.e. decreasing ν_2 with increasing S_4, or positive, i.e. increasing ν_2 for increasing S_4, respectively. All species except for S_1 are inside the established negative or positive feedback loop, they are hence referred to as inner-loop species as well as reactions ν_4 and ν_6 which are referred to as inner-loop reactions.

Both a negative and a positive action of S_4 on ν_2 enable the occurrence of sustained oscillations. As already mentioned, for the occurrence of oscillations, a negative feedback loop is necessary which brings the system back to its state of the beginning. In the negative feedback chain model, a negative feedback is explicitly included by the negative action of S_4 on ν_2 establishing a negative feedback loop comprising S_2, S_3 and S_4. For the chain model with positive feedback, the negative feedback loop is established through the action of ν_2 on its source species S_1. A positive action of S_4 on ν_2 enhances the production of the second, third and fourth species thereby forming a positive feedback loop. But at the same time, enlarging reaction rate ν_2 for increasing S_4 establishes a negative effect on the concentration of S_1 which is reduced more effectively through the conversion to the second species. This negative interaction of S_4 on S_1 establishes a negative feedback loop comprising all four species and makes oscillations possible despite the positive feedback only. Note that similarly, the negative feedback chain model encompasses a positive feedback loop wrapped around the negative feedback loop.

Hence, both chain models yield sustained oscillations, and consequently, the effect of the feedback type on the sensitivity of oscillatory properties can be directly ex-

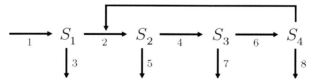

Figure 4.1.: Chain model. The model species $S_1, \ldots, S_4$ form a chain of length four. There are eight reactions, their rates are denoted by $\nu_1, \ldots, \nu_8$, the representing arrows are indexed accordingly. S_1 is assumed to be produced by a constant reaction rate ν_1. All species undergo degradation ($\nu_3, \nu_5, \nu_7, \nu_8$). S_4 acts on the reaction from S_1 to S_2, ν_2, either positively or negatively establishing a positive or negative feedback, respectively.

amined and compared without introducing further reactions or applying any other structural changes to the model. Additionally, the simplicity of the chain models allows to easily implement the two other properties considered in this work - the flow of matter between species (chapter 5) and the type of kinetics (chapter 6).

4.2.1. Implementation of the feedback

The negative and positive action of the last species S_4 in the chain on reaction ν_2 is implemented by a feedback term, fb, entering as a factor into the reaction rate of ν_2. The feedback term of the negative feedback chain model, fb_n, is modeled by an inhibitory Hill kinetics expression depending on S_4 with inhibition constant kn_1 entering as nl-parameter and Hill coefficient n entering as cooperativity parameter and is given by

$$fb_n = \frac{kn_1^n}{kn_1^n + S_4^n}. \tag{4.1}$$

The feedback term of the positive feedback chain model, fb_p, is chosen to be the multiplicative inverse of the negative feedback term, i.e. $fb_p \cdot fb_n = 1$, with activation constant kn_1 and Hill coefficient n:

$$fb_p = 1 + \left(\frac{S_4}{kn_1}\right)^n. \tag{4.2}$$

The choice of the feedback term is discussed in section 4.5.2. The reaction rate ν_2 which is altered by the feedback is changed by multiplying the feedback term and the original reaction rate, e.g. $\nu_2 = k_2 \cdot S_1 \cdot fb_n$ for the chain model with mass action kinetics and negative feedback. The feedback term is completely determined by the ratio $\frac{S_4}{kn_1}$ together with the Hill coefficient. Since during the sampling procedure, the steady state values and the nl-parameters, hence also S_4 and kn_1, are sampled in the interval $(10^{-3}, 10^3)$, their species-to-constant ratio $\frac{S_4}{kn_1}$ takes values in the interval $(10^{-6}, 10^6)$ (compare Figure A.4 C in the Appendix, section A.1.4).

Similar to the notion of feedback, also the feedback strength of the feedback is defined inconsistently in literature. It may be referred to as the inverse to the inhibition or activation constant, $1/kn_1$ (e.g. Qu and Vondriska [2009], Nguyen [2012]), the value of the feedback term at the specific concentration (typically in case of steady state, e.g. Xu and Qu [2012]) or the rate coefficient of a reaction within the feedback loop (e.g. Tian et al. [2009], Zamora-Sillero et al. [2011]). In this work, however, the feedback strength ξ is defined as the percentaged change of the reaction rate which includes the feedback term with respect to changes of the species exerting the feedback (e.g. Steuer [2007]). This coincides with the elasticity coefficient of MCA of the regulated reaction for the regulating species. It considers the concentration of the regulating species, the activation or inhibition constant and the Hill coefficient. All three entities influence the strength of the coupling between the species and thus the feedback strength, although also the rates of the other reactions participating in the feedback loop matter for the actual strength of the coupling between the species and thus the feedback strength. For the chain models, an advantage of the choice for the feedback strength in this work over other definitions from literature becomes clear in connection with the stability analysis in the following section 4.2.2.

In the case of the chain models, the feedback strength of the negative and positive feedback is given by

$$\xi = \frac{\partial log(\nu_2)}{\partial log(S_4)} = \frac{S_4}{\nu_2} \frac{\partial \nu_2}{\partial S_4}. \tag{4.3}$$

Since the feedback term fb_p is the multiplicative inverse of the feedback term fb_n and $log(x^{-1}) = -log(x)$, also the formulas for the norms of the partial logarithmic derivatives are the same. This leads to

$$|\xi_n| = |\xi_p| = n \cdot \frac{S_4^n}{kn_1^n + S_4^n}, \tag{4.4}$$

where the feedback strength in the negative feedback chain model, ξ_n, is negative and the feedback strength in the positive feedback chain model, ξ_p, is positive.

The representation of an inhibitory effect of a species on a reaction as given in Equation 4.1 is widely used in a variety of models showing oscillations (e.g. the Goodwin oscillator: Goodwin [1965], the repressilator: Elowitz and Leibler [2000], MAPK models: Kholodenko [2000], circadian oscillations: Locke et al. [2005a], Goldbeter [1995], Becker-Weimann et al. [2004], calcium oscillations: Sneyd and Dufour [2002], NF-κB oscillations: Ashall et al. [2009]). The type of positive action given in Equation 4.2 term resembles a power-law with a constant basal reaction rate. It is only rarely employed in literature. However, three properties make it very suitable for modeling a positive feedback: (i) The feedback term is always larger than one, i.e. the reaction rate where the feedback enters is not depending on the existence of S_4 and indeed is truly increased if the feedback kicks in. (ii) The feedback term can be deliberately high. This is the exact opposite to the negative feedback term which can be arbitrarily close to zero and hence arbitrarily small without changing the direction of the reaction. (iii) The feedback strength as defined in Equation 4.4 of this positive feedback term is exactly the feedback strength of the negative feedback term with reversed sign for the same values of S_4, kn_1 and n. Hence, the positive action of S_4 on ν_2 captures well the opposite behavior to the negative action of S_4 on ν_2.

4.2.2. Stability analysis and fixation of the Hill coefficients

Even if a negative feedback loop is present, sustained oscillations are only feasible if there is a sufficient delay in it (Novak and Tyson [2008], Stricker et al. [2008], Hogenesch and Ueda [2011]). Since the length of the chain is fixed to four, there is no variability in the number of processing steps to tune the delay. To obtain a sufficiently strong delay, it remains the possibility to alter the cooperativity (Wagner [2005]) by altering the value of the Hill coefficient.

As described in section 2.3, the approach used here is only able to detect sustained oscillations if the ODE system possesses an unstable steady state for the parameter set in question. Therefore, for each of the two chain models, a stability analysis is performed. The percentage of sets with a stable steady state obtained during the parameter sampling procedure is depicted in color-code in dependence on the Hill coefficient and the ratio $\frac{S_4}{kn_1}$ (Figure 4.2) A and B for the negative feedback, C and D for the positive feedback chain model).

All in all, it can be observed that the chain model with negative feedback (Figure 4.2 A) exhibits far more often stable steady states than the chain model with positive feedback (panel B). This is as expected since it is known that a negative feedback reduces noise and hence stabilizes the steady state. However, the observation does not automatically imply that also sustained oscillations occur more frequently for the positive than for the negative feedback chain model. Instead, a positive feedback enhances bistability where two stable steady states arise from one by parameter change, and where the originally unique stable steady state turns unstable. Figure 4.2 B and D show the results of a stability analysis with larger resolution in $\frac{S_4}{kn_1}$ close to values of one. For both feedback types, these panels reveal an abrupt change in the stability behavior for the species-to-constant ratio being close to one. Both chain models do not exhibit unstable steady states for parameter sets having a species-to-constant-ratio below 0.7. For higher species-to-constant-ratios, unstable steady states occur. The underlying reason is the change of feedback strength in the models (as defined in Equation 4.4). The white level lines in Figure 4.2 B and D illustrate the development of the feedback strength in the course of the stability

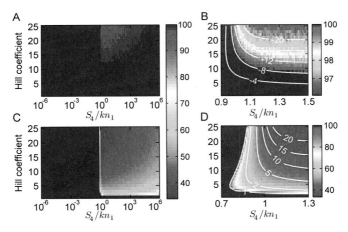

Figure 4.2.: Stability of the chain models with negative or positive feedback. S_4/kn_1 determines the size of the feedback term in case of negative and positive feedback. A, C: 10^4 parameter sets are sampled for each combination of an integer Hill coefficient in $[1, 25]$ and each species-to-constant-ratio in one of the intervals $\{(10^{-6}, 10^{-5.9}), \ldots, (10^{5.9}, 10^6)\}$ for the chain model with either negative (A) or positive feedback (C), and the eigenvalues of the Jacobian matrix for the sampled steady state are determined. Indicated is the percentage of parameter sets exhibiting stable steady states meaning that all real parts of the eigenvalues are negative (section 2.3.2). B, D: Results of the same analysis for each integer Hill coefficients in $[1, 25]$ and S_4/kn_1 in bins of width 0.01 close to one together with the level curves (white) for the feedback strength as defined in Equation 4.4.

analysis. They demonstrate that for the negative feedback (panel B), the absolute feedback strength needs to be larger than eight in order to obtain instabilities whereas for the positive feedback (panel D), a feedback strength of greater than one suffices. Additionally, the percentage of unstable steady states is directly correlated to the feedback strength, i.e. the stronger the feedback strength the more frequently unstable steady states are observed. This is a clear advantage of the definition of the feedback strength employed here since other measures do not display this concordance (compare the Appendix, section A.2.1).

The minimal feedback strength which is necessary to obtain unstable steady states leads to a suitable fixation of the Hill coefficients for the chain models in order to observe sustained oscillations during the sensitivity analysis. The maximal value of the feedback strength for a fixed Hill coefficient corresponds to the value of this Hill coefficient since for non-negative inhibition or activation constant kn_1, the second factor in Equation 4.4 is always smaller than or equal to one. This means that the possible feedback strength is larger the larger the Hill coefficient of the feedback term is. For each feedback type, the smallest possible Hill coefficient is chosen which enables the occurrence of unstable steady states. Recall that feedback strengths greater than eight or one are necessary for the negative or positive feedback chain model, respectively. Therefore, a value of $n = 9$ is set as Hill coefficient for the negative feedback chain model (in accordance with findings from Griffith [1968a] and Nguyen [2012]) and a value of $n = 2$ for the positive feedback chain model. Other Hill coefficients for the chain models and their effects on the sensitivities are discussed in section 4.5.2. The Hill coefficient, which is the cooperativity of the feedback term, is considered to be a structural property of the system and is hence not altered in the sampling process.

4.3. Sensitivity analysis of the chain model with positive or negative feedback

The linear chain models have been introduced by Wolf et al. [2005] and therein analyzed with respect to period sensitivity for each feedback type for one particular parameter set for which the system yields sustained oscillations. For this reference parameter set, the negative feedback chain model was found to exhibit lower period sensitivity than the positive feedback chain model. This finding hinted towards the conclusion that negative feedback causes rather low period sensitivities, but it was not shown to which extent the choice of the parameter set is responsible for this. To clarify this question, a sensitivity analysis for the chain models of with random sampling for the parameter sets as explained in section 2.3 is performed. The results of the sensitivity analysis are given in Figure 4.3 and in the Appendix (Tables A.8, A.9). In the scatter plot of Figure 4.3, the orange stars denote the sensitivities for the chain model with negative feedback, the green squares those for the positive feedback chain model. Black symbols show the median values, the inset depicts the sensitivity values larger than 50. The box-plots in the according colors to the sides of the scatter plot capture the sensitivity distribution characteristics.

The sensitivities of the positive and negative feedback chain model segregate into

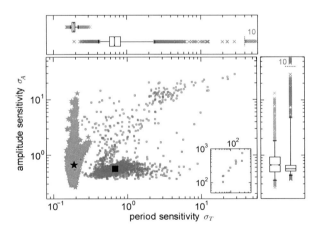

Figure 4.3.: Impact of the feedback type on the sensitivity of the chain models. Amplitude and period sensitivities for the negative feedback chain model (orange stars, black star: median values) and positive feedback chain model (green squares, black square: median values) governed by mass action kinetics. Sensitivities are computed as in Figure 3.2. The inset displays parameter sets with sensitivity values > 50. The box-plots show the characteristics of the sensitivity distributions for period sensitivity (top) and amplitude sensitivity (right) in the according colors.

two different populations with only very minor overlap. This is mainly due to the period sensitivity distributions. Compared to the positive feedback chain model, the period sensitivities of the negative feedback chain model are significantly smaller (MWU p-value: 0, Table A.9), the median values being 0.19 and 0.68 for the positive and negative feedback chain model, respectively. Even the intervals formed by the fifth and 95th quartiles are mutually exclusive ([0.17, 0.22] and [0.41, 2.40] for negative and positive feedback chain model, respectively, Table A.8). For the negative feedback chain model, the width of the period sensitivity distribution measured by the 90%R is nearly 40-fold less than that of the positive feedback chain model (90%R: 0.05 and 1.99, respectively, Table A.8). This means that the period sensitivity is not only lower for the negative feedback chain model, but it is also decisively less influenced by the choice of the parameter set than the period of the positive feedback chain model. One may conclude that a negative feedback intrinsically leads to low whereas a positive feedback to higher period sensitivities.

For the amplitude sensitivity distributions, the tendency reveals to be rather converse. The median amplitude sensitivity for the positive feedback chain model takes a value of 0.57 and is lower than that of the negative feedback chain model being 0.66 (Table A.8). The amplitude sensitivity distributions show an overlap on the level of the first and third quartiles ([0.50, 0.93] for the negative feedback, [0.51, 0.65] for the positive feedback, Table A.8). The difference between the distributions indicated by the median values is highly significant (MWU p-value: $5 \cdot 10^{-24}$, Table A.9) although the amplitude sensitivities reach the lowest values for the negative feedback chain model and the highest for the positive feedback chain model. This means that the positive feedback chain model mainly exhibits lower amplitude sensitivities than does the negative feedback chain model. Concerning the width of the distributions, on the level of the quartiles, the positive feedback chain model yields a three-fold smaller distance between first and third quartile than the negative feedback chain model, for the 90%R, it is exactly opposite (90%R: 1.49 and 4.48 for the negative and positive feedback chain model, respectively, Table A.8). Hence, compared to the differences in distribution width for the period sensitivities, the distributions widths for the amplitude sensitivities are rather similar for the two chain models.

Also, both models yield similar, positive but low correlations between their period sensitivities and amplitude sensitivities ($R_S = 0.3$ and $R_S = 0.42$ for negative and positive feedback, respectively). This indicates that by the choice of the parameter set, period sensitivity and amplitude sensitivity can be tuned rather independently from each other within the limits imposed by the according chain model structure.

Summing up, for the negative feedback chain model in general, very low period sensitivities are observed while experiencing variable amplitude sensitivities. For the positive feedback chain model, the parameter sets in the middle 50% of the amplitude distribution display highly conserved, low amplitude sensitivities and variable period sensitivities higher than for the negative feedback chain model.

4.4. Further investigation of the sensitivities of the chain models

4.4.1. Branching in the period sensitivity of the negative feedback chain model

For the sensitivities of the negative feedback chain model (orange stars in Figure 4.3), two branches appear: The majority of parameter sets (as indicated by the median) is situated on the left, it displays a distribution with stable, low period sensitivity (lower than 0.213) and varying amplitude sensitivity. To the right, a smaller amount of parameter sets form a second branch where the period sensitivity is enlarged and the amplitude sensitivity increases with increasing period sensitivity. A further analysis of the parameter sets in this right branch shows that an underlying reason for this qualitatively different behavior could be the maximal flow distribution in the system.

All parameter sets for the chain model with negative feedback with a period sensitivity larger than 0.213, which includes all parameter sets of the right branch, share the property of having a larger maximal conversion flow from S_1 to S_2 (modeled by ν_2) than degradation rate of S_1 (modeled by ν_3), i.e. $\max(\nu_2) > \max(\nu_3)$. $\max(\nu_i)$ hereby is the maximal possible value of the flow obtained during an oscillation, calculated for maximal (resp. minimal, if applicable) species concentration. In Figure 4.4 A, the sensitivities of all 555 parameter sets with $\max(\nu_2) > \max(\nu_3)$ are depicted by dark orange dots, together with the sensitivities of the remaining 1945

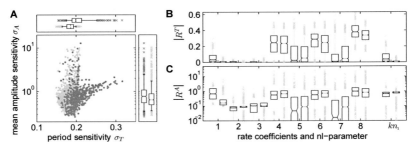

Figure 4.4.: Different sensitivities in the negative feedback chain model. The 2500 parameter sets of the negative feedback chain model are divided into two groups: those with $\max(\nu_2) > \max(\nu_3)$ (dark orange, 555 sets) and those with $\max(\nu_2) \leq \max(\nu_3)$ (light orange, 1945 sets). A: The results of the sensitivity analysis for both groups. B and C: Box-plots of the absolute sensitivity coefficients for each parameter for period sensitivity (B) and amplitude sensitivity (C). The two groups are marked by the color of their medians (left: dark orange, $\max(\nu_2) > \max(\nu_3)$, right: light orange, $\max(\nu_2) \leq \max(\nu_3)$).

parameter sets with $\max(\nu_2) \leq \max(\nu_3)$ (light orange). The box-plots in Figure 4.4 A indicate that also the median period sensitivity and median amplitude sensitivity of the 555 parameter sets with larger maximal ν_2 than ν_3 are larger.

The reason for this altered sensitivity in dependence on the maximal flow distribution could be explained by the presence of different paths of degradation of S_1 whose degradation is crucial for oscillations to arise. In the chain model, S_1 is directly degraded via reaction ν_3. Alternatively, there is a decrease of S_1 by conversion to S_2 via reaction ν_2 and further processing through the chain and degradation of the other, inner-loop species. Whenever ν_3 is small compared to ν_2, this alternative degradation path constitutes the major part of the decrease of S_1. Thus, the sensitivity of the period and amplitude towards certain parameter changes is altered. This is shown by the distribution of the period sensitivity coefficients which are given in Figure 4.4 B. Therein, for each kinetic parameter, the values for the absolute period sensitivity coefficients for the 555 parameter sets with $\max(\nu_2) > \max(\nu_3)$ (dark orange) and for the remaining 1945 parameter sets with $\max(\nu_2) \leq \max(\nu_3)$ (light orange) are given. The plot evidences that having $\max(\nu_2) > \max(\nu_3)$ causes an increase in the absolute period sensitivity coefficients: (i) of the inhibition constant

kn_1 (which tunes how fast S_1 is converted to S_2), (ii) of the production rate of S_1 (rate coefficient 1, alteration in there is strongly affecting the production of species S_2 for large $\max(\nu_2)$), (iii) of the conversion from S_3 to S_4 (rate coefficient 6) and (iv) of the degradation of S_4 (rate coefficient 8) which constitute the last part of the alternative degradation path of S_1. At the same time, only rate coefficients 5 and 7 determining the degradation rates of species S_2 and S_3 exhibit slightly lower period sensitivity coefficients for the parameter sets with $\max(\nu_2) > \max(\nu_3)$. The sensitivity towards changes of other kinetic parameters remains unchanged. Altogether, this can lead to the observed increase of the overall period sensitivities σ_T.

In Figure 4.4 C, a potential source of the increased median amplitude sensitivity for the parameter sets with $\max(\nu_2) > \max(\nu_3)$ is revealed: The absolute amplitude sensitivity coefficients of rate coefficient 1 determining the constant production of S_1 are decisively enlarged, whereas only the already very small sensitivity coefficients of rate coefficients 5 and 7 (governing the degradation rates of species S_2 and S_3, respectively) are further reduced. Indeed, this stronger impact of alterations of the production rate of S_1 on amplitude sensitivity for the parameter sets with $\max(\nu_2) > \max(\nu_3)$ is as expected. For these parameter sets, alterations in the production of S_1 are difficult to compensate without affecting also species S_2 due to the small degradation rate ν_3 of S_1 compared to the conversion rate to S_2, ν_2. This leads to stronger variations of the amplitude of S_1 which can additionally be more easily propagated to the other species of the chain.

Still, the differences of the period sensitivity between the two branches are very low compared to the differences between the negative and the positive feedback chain model.

4.4.2. Bifurcation analyses of the negative feedback chain model

To further examine the origin of the low period sensitivity and variable amplitude sensitivity found in section 4.3, a bifurcation analysis (see section 2.1.7) is performed for some of the parameter sets of the negative feedback chain model yielding different sensitivity values (Figure 4.5 D). As bifurcation parameter, the parameter yielding the largest absolute period sensitivity coefficient is used. For a parameter set with sensitivity values close to the median values (parameter set A indicated in panel D), parameter k_6 governing the conversion reaction from S_3 to S_4 exhibits the largest period sensitivity coefficient as well as amplitude sensitivity coefficient. The bifurcation diagrams for this parameter are given in Figure 4.5 A. For low values of k_6, the system has a stable steady state (red line). By increasing k_6, the system passes through a supercritical HB and stable LCs (green dots) occur, the steady state turns unstable (dotted black line). Further increasing k_6 then leads to another supercritical HB where the LCs vanish and the steady state turns stable again.

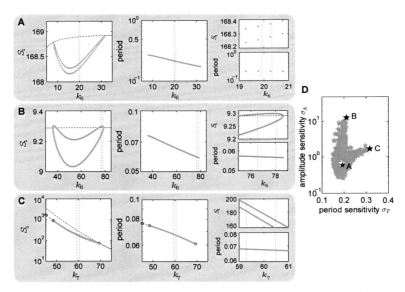

Figure 4.5.: Bifurcation diagrams for the chain model with negative feedback. For each of the parameter sets marked by a black star in panel D, a bifurcation analysis is performed for the parameter with largest absolute period sensitivity coefficient. Given are plots of species S_1 or the period versus the bifurcation parameter. Small panels show details for specific intervals of the bifurcation parameter. Red lines denote stable steady states, dotted black lines unstable steady states. Green dots (often melting to a line) denote stable LCs (pair on vertical lines give maximal and minimal value for the specified concentration), blue circles unstable LCs. The dotted gray vertical lines indicate the original value of the bifurcation parameter and its perturbation (+2%) for the parameter set examined.

The amplitude varies smoothly, being low at the bifurcations and higher the further away from the bifurcations. It is varying more the closer to the HB. The actual value of k_6 for this parameter set is relatively far from the bifurcations (gray vertical lines for the original and the value perturbed by $+2\%$); therefore, the change in amplitude is rather low (amplitude sensitivity coefficient for k_6: $R_{k_6}^A = -0.96$). For the whole range where stable LCs occur, the period is changing only slightly and smoothly with changing k_6. In contrast to the amplitude, the distance of the parameter value from the bifurcations seems not to influence how large the changes in period are.

For the parameter set with largest amplitude sensitivity (parameter set B, Figure 4.3 B), similar bifurcation diagrams for k_6 as in A are obtained. The increased amplitude sensitivity coefficient (here taking a value of -24.2) arises from the specific value of k_6 being very close to one of the HBs where the amplitude strongly varies. Over the whole range of the occurrence of stable LCs, the period changes only slightly which results in a low period sensitivity coefficient.

For the parameter set with the largest period sensitivity (marked by C in Figure 4.5 D), the bifurcation diagram for k_7 as parameter with largest period sensitivity coefficients also displays the occurrence of unstable LCs (blue circles, Figure 4.5 C). Note that the maximal and minimal values of S_1 are close to each other compared to their distance to the unstable steady state and therefore only one green or blue dot is visible for each parameter value. Nevertheless, the system passes from a stable steady state through a supercritical HB for decreasing k_7 and still stable LCs occur. The characteristics of the change in amplitude and period remain similar to those obtained for the other two parameter sets: The amplitude varies more strongly the closer the parameter is to the HB, the period varies only slightly overall.

These bifurcation analyses support the view that the period may be intrinsically only slightly variable for the negative feedback chain model, irrespective of the particular choice of the parameter values and hence parameter set. In contrast, the amplitude varies more strongly the closer the parameters are to the HBs. This may be the reason for the variable amplitude sensitivities occurring in the negative feedback chain model (Figure 4.3).

4.4.3. Bifurcation analyses and birhythmicity in the positive feedback chain model

The sensitivities of the positive feedback chain model can reach up to very high values. In fact, for the parameter sets with high sensitivities, it is found that often the sensitivity coefficient of one single or only few parameters are very high, indicating an extreme alteration in period and/or amplitude for slight parameter variations. To further examine into the origin of this behavior, bifurcation analyses for the parameter with highest absolute period sensitivity coefficient of some parameter sets with different sensitivities is performed. The results are shown in Figure 4.6.

First of all, a parameter set with period sensitivity and amplitude sensitivity close to the median values (marked by A in Figure 4.6 E) is examined. In Figure 4.6 A, the results of the bifurcation analysis for the inhibition constant kn_1 is shown. For increasing kn_1, the stable steady state turns unstable and then stable again. The system passes twice through supercritical HBs where stable LCs arise. The details of the system behavior around the specific parameter value of kn_1 and its perturbation by $+2\%$ (gray dotted lines) are given in the first two small panels. Only stable LCs occur and neither the period nor the amplitude of S_1 change dramatically when changing the parameter which corresponds to the parameter set exhibiting only median period and amplitude sensitivity. Between the two HBs, two regions are observed where unstable LCs occur (left panel in Figure 4.6 A). Details of the bifurcation diagram around the according parameter values are shown in the four plots to the right. The unstable LCs between $kn_1 = 0.001828$ and $kn_1 = 0.001831$ separate two stable limit cycles with very different periods and amplitudes which coexist for these parameter values. This means that for the chain model with positive feedback, birhythmicity can be observed there, and a parameter change can lead to the switch from one LC to the other. The unstable LCs close to the right HB separate the arising stable LCs from both HBs. There, many bifurcations are detected in a very narrow interval of parameter kn_1. If stepping over this interval by a parameter perturbation (as e.g. in the sensitivity analysis), a step-like, steep change of period and amplitude is experienced. This means that if choosing another value for kn_1 (close to the left of the unstable LCs), enhanced sensitivities for changes in kn_1 instead of the actual intermediate sensitivities can arise.

In Figure 4.6 B, the parameter sets marked by B in Figure 4.6 E is examined. This parameter set exhibits higher period sensitivity than the median value but median amplitude sensitivity. The bifurcation diagram for kn_1 resembles that of

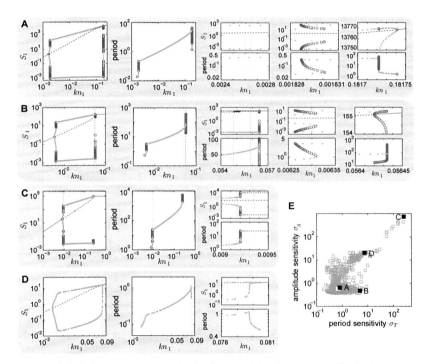

Figure 4.6.: Bifurcation diagrams for the chain model with positive feedback. For each of the parameter sets marked by a black square in panel E, a bifurcation analysis is performed for the parameter with largest absolute period sensitivity coefficient. Given are plots of species S_1 and the period versus the bifurcation parameter. Small panels show details for specific intervals of the bifurcation parameter. Red lines denote stable steady states, dotted black lines unstable steady states. Green dots (often melting to a line) denote stable LCs (pair on vertical lines give maximal and minimal value for the LC), blue circles unstable LCs. The dotted gray vertical lines indicate the original value of the bifurcation parameter and its perturbation ($+2\%$) for the parameter set examined.

Figure 4.6 A. The difference is that the upper HB is a subcritical instead of a supercritical one (to be seen in the small panels to the right). This bifurcation causes the occurrence of the unstable LCs near the upper limit of kn_1 for oscillations. Indeed, the specific parameter value of kn_1 is close to this HB where the change of the period is enhanced compared to the rest of the regions with stable LCs (steeper slope of period change in the plot of period vs kn_1) leading to an increased period sensitivity coefficient $R^T_{kn_1} = 115.0$. In contrast, the amplitude is not considerably altered by parameter perturbation delivering a reason for the observed high period but only median amplitude sensitivity of the parameter set. The middle detailed plots show that in the region to the left with unstable LCs, the system exhibits again birhythmicity.

The results of a bifurcation analysis of the parameter set with highest period and amplitude sensitivity (marked by C in Figure 4.6 E) are given in Figure 4.6 C. Again, for increasing kn_1, a supercritical HB leads to the occurrence of stable LCs. For even larger kn_1, the system passes through a subcritical HB where the stable steady state reappears. For the perturbations of the specific parameter value of kn_1 (detailed plots on the right), the system steps from one regime of stable LCs to another with strongly increased period and amplitude. Therefore, the sensitivity coefficients for kn_1 are very high taking values of $R^T_{kn_1} = 380$ and $R^A_{kn_1} = 1104$. Though, birhythmicity cannot be detected in this region. If existing, its interval of occurrence for kn_1 is smaller than $3 \cdot 10^{-8}$. The overall sensitivities σ_T and σ_A at this parameter set are only very high because the same type of step-like increase in period and amplitude also occurs for the parameters k_5, k_7 and k_8 (data not shown) determining the degradation rates of S_2, S_3 and S_4.

For some of the parameter sets, even though high sensitivity is observed, no unstable LCs and hence only smooth changes in amplitude or period are existent. This is the case for the parameter set marked by D in Figure 4.6 E whose bifurcation diagrams for kn_1 are shown in Figure 4.6 D. The HBs are both supercritical leading directly to stable LCs. Being close to the upper HB, at the specific parameter value, high period and amplitude sensitivity coefficients are obtained.

For all parameter sets examined with this bifurcation analysis, the shapes of the curves in the bifurcation diagrams resemble each other. The regions where the amplitudes remain nearly constant are very large compared to the tiny regions close to the HBs where the amplitudes change strongly. In contrast, the changes in periods are getting steeper for parameter values getting higher, not necessarily close to the HBs. This may be the reason why the majority of parameter sets for the chain model with positive feedback exhibit low, constant amplitude sensitivities but varying period sensitivities as observed in Figure 4.3. Parameter sets with increased amplitude and period sensitivity may arise by one or more parameter values being close to the HBs or by one or more parameters passing a step-like period and amplitude change. The latter may be caused by birhythmicity phenomena.

4.5. Summary and discussion: The effect of feedbacks on sensitivity

In this chapter, the chain models of length four with mass action kinetics are introduced as prototype oscillator models (section 4.2). They yield sustained oscillations for a negative as well as a positive feedback. Hence, a direct comparison of the effect of negative or positive feedback types on the period and amplitude sensitivities without the need to introduce further reactions, regulations or parameters is enabled. Additionally, also the other structural properties (matter flow, kinetics) which will be examined in the next chapters can easily be implemented and their effect can be observed in dependence on the feedback type present. Note the difference in the notion feedback and feedback loop as discussed in section 4.1.

A sensitivity analysis shows that for the two different feedback types, the chain models exhibits different period and amplitude sensitivity distributions if analyzing a large number of randomly sampled parameter sets (section 4.3). Thereby, if including a negative feedback only, the period sensitivity is low for every parameter set and varies only in a narrow range while the amplitude sensitivity varies over a broad range and takes higher values. Slight increases in period sensitivity might be credited to an increased flow through the chain as opposed to a fast degradation of S_1 (section 4.4.1). Bifurcation analyses of particular parameter sets indicate that the amplitude sensitivity coefficients may be increased if the parameter values are close to the HB whereas the period sensitivity coefficients remain low without being influenced by the specific parameter value (section 4.4.2).

If exhibiting a positive feedback, the majority of sensitivity values obtained for the chain model cluster in a region with low and only slightly varying amplitude sensitivities and broadly varying period sensitivities. Very high period and amplitude sensitivities can also occur. Bifurcation analyses of particular parameter sets reveal the occurrence of steep changes of the period close to HBs (section 4.4.3). In contrast to the negative feedback chain model, parameter sets with parameter values a bit further from but still close to HBs do not necessarily reveal steep amplitude changes. These two observations might explain the observation of low, constant amplitude sensitivities and varying period sensitivities for the majority of parameter sets. Additionally, the bifurcation analyses reveal the occurrence of birhythmicity. (section 4.4.3) Hence, step-like changes in the period as well as the amplitude even for slight parameter changes are enabled which result in parameter sets with decisively enlarged period and/or amplitude sensitivity.

4.5.1. Relation to other findings

Findings of the influence of feedback on period and amplitude sensitivity

Instead of sensitivity coefficients, Zamora-Sillero and colleagues [2011] employ as measure of robustness the occupancy of a viable region and the length of a random walk to leave the viable region. Thereby, the occupation of a viable region consisting of parameter sets exhibiting oscillations with the period in a restricted interval is a mixture of oscillation probability *per se* (as every oscillating parameter set can be in principle scaled to exhibit any desired period) and period sensitivity as employed in this thesis. Also the length of the random walk until the viable region is left is closely related to the herein employed period sensitivities as the walk also considers how the period of the oscillations reacts with respect to parameter changes.With respect to these measures and period sensitivity, the results found here and by Zamora-Sillero and colleagues [2011] coincide: Negative feedback architectures have the most robust period. However, in this thesis, positive feedback does *not* increase the robustness of the period, as has been suggested by Zamora-Sillero and colleagues [2011]. The issue of oscillation probability and feedback type is discussed in section 4.5.3.

The results on the tunability of the period and amplitude in negative feedback or negative-plus-positive feedback architectures (Tsai et al. [2008], Stricker et al. [2008])

are confirmed using the sensitivity analysis in this thesis and the chain models with negative or positive feedback, respectively. Here, the negative feedback chain model has been shown to exhibit very low period sensitivities and potentially high amplitude sensitivities. This is in perfect agreement with the finding of a less tunable period (Stricker et al. [2008]) or only minor changes in period while experiencing varying amplitudes for negative feedback only oscillator models (Tsai et al. [2008]). Additionally, the herein found increased period sensitivity of the positive feedback chain model complies with findings of positive feedback loops allowing for greater variability of the period (Stricker et al. [2008]). Even more, the majority of parameter sets of the positive feedback chain model, at least the middle 50% with respect to the amplitude sensitivity, span a large range of period sensitivities while having a rather restricted amplitude sensitivity. This accords with results from Tsai and colleagues [2008] stating the occurrence of a tunable period and a region of constant amplitude if adding a positive feedback loop to their oscillator model.

Obtaining these similar results is striking because the approach in Tsai et al. [2008], Stricker et al. [2008] to assess the sensitivities in period and amplitude is different to that employed in this thesis. The biggest difference is the perturbation of the parameter. In this work, the effect of small changes of all parameters on period and amplitude is applied, instead of the change of one parameter over a broad range. The first avoids bias towards which parameter change is investigated, but also includes the effect of parameters which may be potentially less important. Furthermore, in contrast to the approach with the chain models in this thesis, the amplitude in the work by Tsai et al. [2008] is bound by one since every species stems from a conversion cycle and is modeled as the fraction of active protein. The lack of finding large variance in the amplitude in Tsai et al. [2008] as opposed to parameter sets with very high amplitude sensitivities in the positive feedback chain model might arise due to fact that the species concentrations in the chain model in this thesis are not restricted. Additionally, in Tsai et al. [2008], the one parameter for which the period and amplitude variability has been examined has been varied smoothly. The smooth variation makes it impossible to step over the discontinuities experienced in the positive feedback chain model (section 4.4.3) and therefore leaves a big region unexplored where still oscillations occur but where the period and/or amplitude are decisively altered. Hence, Tsai and colleagues [2008] might have underestimated the variation in amplitude and period. The observed lower amplitude sensitivities in the model with positive feedback in Tsai et al. [2008] might also arise due to the positive feedback being an autoregulatory feedback without intermediate species, the delay of the positive feedback is very short. This might result even in lacking regions of birhythmicity, where step-like changes in concentrations and/or period potentially enable high sensitivities in the chain model with positive feedback.

Overall, quite similar conclusions can be drawn from this thesis and from Tsai et al. [2008] and Stricker et al. [2008] despite the different approaches. This shows that further general principles and their influence on period and amplitude sensitivity might be predictable with the method and models employed in this work.

Relation to the calcium and circadian oscillation models

Interestingly, the findings for the low period sensitivities of the circadian and the high period sensitivities of the calcium oscillations model in Figure 3.2 are reproduced by the negative and positive feedback chain model, respectively. This supports the view that circadian oscillations might rather rely on a negative feedback, whereas the dominating feedback for calcium oscillations might be a positive one.

Additionally, in the positive feedback chain model, with the help of bifurcation analyses, regions of birhythmicity, i.e. the coexistence between two stable periodic regimes, have been observed. Birhythmicity is reported to occur if two instability-generating mechanism are coupled (Decroly and Goldbeter [1982], Goldbeter [2002]). This is the case for both chain models as they both possess a positive as well as a negative feedback loop. However, only the positive feedback chain model is found to display birhythmic regions. The occurrence of birhythmicity has also been reported in a model of bursting calcium oscillations (Haberichter et al. [2001]). This strengthens the conclusion that calcium oscillations and especially their sensitivity characteristics may be better represented by the positive feedback chain model than by the negative feedback chain model.

However, the calcium oscillations model displays higher amplitude sensitivities than the circadian oscillations model, and *vice versa* for the positive and negative feedback chain models. In fact, mainly oscillations with constant amplitudes have been found in calcium signaling (Grubelnik et al. [2001]), and therefore the conclusion on positive feedbacks being dominating feedbacks in calcium models is also supported by the observations concerning the amplitude sensitivity.

4.5.2. Influence of the model assumptions on sensitivity

Choice of the feedback term

The choice of the feedback term is described in section 4.2.1. The Hill term for inhibition used in this work is also widely employed for models of many oscillatory systems in literature (see detailed list of publications in section 4.2.1). However, the employed activation term for the positive feedback which resembles a power-law activation is found less frequently in literature. In most models, the additive inverse $\frac{S^n}{K_A^n + S^n}$ of the inhibitory Hill term is used as activation term (circadian oscillations: Becker-Weimann et al. [2004], Goldbeter et al. [1990], Locke et al. [2005b], calcium oscillations: De Young and Keizer [1992]). To estimate the effect of the one choice over the other for the activation term, a sensitivity analysis for the chain model using the above activation term is performed (Appendix, section A.2.2). It reveals very high period and amplitude sensitivities compared to the original positive feedback chain model. Thereby, it reproduces the sensitivity results found for the calcium oscillations model. At the same time, it does not reproduce the finding of low amplitude sensitivities for the chain model with the original positive feedback term and Tsai et al. [2008] for their model including a positive feedback. It would be interesting to also examine this implementation of activation further, but it yields decisively less oscillations in the chain model and hence requires a computationally more intensive study. Therefore, although less frequently employed in literature, this work restricts further examination of the chain model to the originally defined positive feedback term, which also constitutes a good choice for the three reasons explained in section 4.2.1: The feedback term is always larger than one, can deliberately increase, and yields, up to the sign, the same feedback strength as the negative feedback term.

Hill coefficient

With the help of a stability analysis, the minimal Hill coefficients necessary to enable oscillations in the chain models are determined to be $n = 9$ for the negative feedback and $n = 2$ for the positive feedback. The positive feedback chain model with a Hill coefficient of $n = 9$ as in the negative feedback chain model is analyzed (Appendix, section A.2.2). For this enlarged Hill coefficient, even more parameter sets cluster in a region with low and hardly variable amplitude sensitivity but a broad range of rather large period sensitivities. Although having a larger overlap of the sensitivity values with the negative feedback chain model, the tendency of lower period sensitivities but higher amplitude sensitivities of the negative feedback chain model compared to the positive feedback chain model persists. Therefore, one can conclude that the differences in the sensitivity distributions obtained for the chain models with different feedback types do not originate from the different values of the Hill coefficient.

Length of the feedback loops

In the chain models, it is not only the feedback type which is different, but especially the length of the negative and positive feedback loops established by the feedback. As described in the introduction of the chain models (section 4.2), the negative feedback loop for the negative feedback chain model consists of S_2, S_3 and S_4 whereas that for the positive feedback chain model comprises all four species, and *vice versa* for the positive feedback loop. With increasing chain length, also the delay of the negative feedback loop increases. Delay in the negative feedback loop is necessary for oscillations as discussed in Wagner [2005], Novak and Tyson [2008], Stricker et al. [2008], Hogenesch and Ueda [2011]. Single negative feedback loops comprising less than three species are not yielding oscillations (Goodwin [1965], Novak and Tyson [2008]), hence four is the minimal chain length for the negative feedback chain model to generate oscillations. Sensitivity analyses for chain models of length five with positive or negative feedback of species S_5 on reaction ν_2 have been performed (Appendix, section A.2.3). All in all, prolonging the chain does not alter the sensitivity properties of the models decisively, the negative feedback chain models yield similar sensitivities, and also the positive feedback chain models of different lengths. Therefore, although having exactly the same length of the negative feedback loop (consisting of four species), the positive feedback chain model of length four and the negative feedback chain model of length five do not yield similar sensitivity values. Analogously, for the two models yielding a length of four for the positive feedback loop (negative feedback chain model of length four and positive feedback chain model of length five), the sensitivities are not similar. One can conclude that indeed the feedback type applied determines the sensitivity behavior and hence the order of the feedback loops, i.e. whether the positive feedback loop is inside or outside the negative feedback loop. The thereby resulting lengths of the negative or positive feedback loops seem to play only a minor role.

4.5.3. Oscillation probability

With respect to the oscillation probability, Stricker and colleagues [2008] propose that oscillations occur for a large range of inducer concentration if a positive feedback is present. Accordingly, for the stochastic model with only a negative feedback, the oscillations are observed to get less regular, i.e. the period and amplitudes are more variable during the oscillations, and oscillations are observed in a smaller parameter range in the deterministic model. Also Tsai and colleagues [2008] state that their model which includes a positive feedback shows more often oscillations than the model with negative feedback only. These results are also found for the chain models here. While the oscillation probability (i.e. 2500 divided by the number of sampled parameter sets, Table A.8) is 0.013% for the negative feedback model, it is nearly 130-fold more, i.e. 1.67%, for the positive feedback model.

In Zamora-Sillero et al. [2011], concerning oscillation probability as occupancy of the viable region, the results are only similar to those from the examination in this thesis and in Tsai et al. [2008] with respect to an additional positive feedback increasing the oscillations probability. In contrast, Tsai and colleagues [2008] and this thesis find that negative feedback only models oscillate only poorly, and thus they are less robust concerning oscillation probability, while Zamora-Sillero and colleagues [2011] consider the essential negative-feedback architectures most robust. Probably, the definition of the viable region as parameter sets "exhibiting oscillations *in the interval* $[0.9, 1.1]$" instead of only "exhibiting oscillations" causes these differences observed for the oscillation probabilities in the different approaches.

The question remains whether the differences in oscillation probabilities between the two chain models are indeed caused by the feedback type and not by some other reasons. A potential source of the increased oscillation probability in the positive feedback chain model may be the longer negative feedback loop present than in the negative feedback chain model (discussed in section 4.5.2 and the Appendix, section A.2.3). Increasing the chain length to five species also increases the length of the negative feedback loop and therefore the delay in it. Delay is necessary for oscillations to occur and it may also affect the oscillation probabilities. The oscillation probabilities for the negative feedback chain model of length five is 0.13% (Appendix, section A.2.3) which is a 10-fold increase in occurring oscillations compared to the negative feedback chain model of length four although the parameter space has been enlarged by the introduction of two additional reactions. On the contrary, prolonging the chain by one species for the positive feedback, the oscillation probability is decreased to 1.04% which is only slightly more than 60% of the value for the positive feedback chain model of length four. Therefore, no general rule for the effect of increasing the delay by increasing the length of the feedback loop on oscillation probability can be derived as the chain models with different feedback types behave differently. Accordingly, the finding of a prolonged negative feedback loop for the positive feedback chain model cannot explain why therein a higher percentage of parameters sets is found to deliver oscillations than in the negative feedback chain model.

Additionally, increasing the feedback strength in the positive feedback chain model decreases the oscillation probability. If applying a Hill coefficient of $n = 9$ instead of $n = 2$ (section A.2.2) and thereby increasing the maximal feedback strength from two to nine, the oscillation probability is 0.25% which is more than six-fold lower than that for the positive feedback chain model with a Hill coefficient of $n = 2$. In this scenario, the feedback strength of the positive feedback present in the system is increased together with the feedback strength of the negative feedback. Increased feedback strength for a negative feedback has been reported to increase the occurrence of oscillations (Nguyen and Kulasiri [2009]). The coupled increase of the positive feedback strength might be the reason for the unexpected decrease in oscillation probability.

4.5.4. More possibilities of feedback implementation

As pointed out in the last section and described in section 4.1, there is not a unique method of establishing a feedback loop via the feedback term in the chain model.

Instead of the direct negative effect of S_4 on the production of S_2 (i.e. on reaction ν_2), a negative feedback loop of the same length could be established if S_4 would have a positive effect on the degradation of S_2 (i.e. on reaction ν_5). However, then turning the positive effect into a negative would lead to a chain model with positive feedback, but without any negative feedbacks thereby not showing oscillations for any parameter set. Hence, this implementation is not possible to use to compare negative and positive feedback without introducing further changes in the structure of the models, i.e. adding reactions or interactions. Therefore, this work is restricted to the original feedback implementation as introduced in the chain model (section 4.2, Wolf et al. [2005]). The question of the effect of feedbacks acting on different reactions in the model on period and amplitude sensitivity remains to be explored as a prospective work.

Furthermore, it has been shown that in the yeast high osmolarity glycerol (HOG) signaling pathway, shorter additional feedbacks help to avoid oscillations (Schaber et al. [2012]). It is possible that this type of shorter feedbacks also affects the period and amplitude sensitivities. This question may be addressed using the chain models and the same approach as undertaken here. However, it is not the focus of this work as it requires to introduce further reactions or regulations and hence parameters into the model and therefore remains to be investigated.

5. The effect of matter flow on sensitivity

In the preceding chapter, the impact of the feedback type on period and amplitude sensitivity has been investigated. Chapter 5 focuses on another property which may influence the period and amplitude sensitivity: Including or lacking matter flow between species. First, the meaning of matter flow between species or lacking matter flow between species for biological processes is clarified (section 5.1). Second, the implementation of matter flow properties in the chain model is described and their effects on the occurrence of oscillations are analyzed (section 5.2). Sensitivity analyses are performed (section 5.3) to reveal the effect of matter flow properties between species for the negative feedback chain model and the positive feedback chain model. Further analyses shed light on the origins of the altered sensitivities in case of lacking matter flow between species (section 5.4).

5.1. Matter flow: conversion vs. regulated production

Matter flow is a property of a biological process which indicates that the number of molecules or the concentration of a species is affected by the process. Whether a process mediates matter flow or not depends on the definition of the species in the system. For example, a transport of a protein from the nucleus to the cytosol does not change the total concentration of the protein in the cell and therefore could be considered as lacking matter flow. However, if considering the cytosolic protein and the nuclear protein separately, both concentrations are altered by the transport which thus mediates matter flow between the two species.

In mathematical models of biological processes, every reaction encoding a biolog-

A pure production

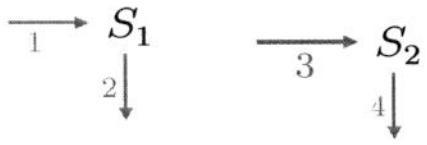

$$\frac{dS_1}{dt} = k_1 - k_2 \cdot S_1$$
$$\frac{dS_2}{dt} = \boxed{k_3} - k_4 \cdot S_2$$

$$\Leftrightarrow \quad \begin{pmatrix} \frac{dS_1}{dt} \\ \frac{dS_2}{dt} \end{pmatrix} = \begin{pmatrix} 1 & -1 & \boxed{0} & 0 \\ 0 & 0 & \boxed{1} & -1 \end{pmatrix} \cdot \begin{pmatrix} k_1 \\ k_2 \cdot S_1 \\ k_3 \\ k_4 \cdot S_2 \end{pmatrix}$$

B conversion

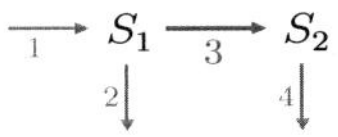

$$\frac{dS_1}{dt} = k_1 - k_2 \cdot S_1 \boxed{- k_3 \cdot f(S_1)}$$
$$\frac{dS_2}{dt} = \boxed{k_3 \cdot f(S_1)} - k_4 \cdot S_2$$

$$\Leftrightarrow \quad \begin{pmatrix} \frac{dS_1}{dt} \\ \frac{dS_2}{dt} \end{pmatrix} = \begin{pmatrix} 1 & -1 & \boxed{-1} & 0 \\ 0 & 0 & \boxed{1} & -1 \end{pmatrix} \cdot \begin{pmatrix} k_1 \\ k_2 \cdot S_1 \\ k_3 \cdot f(S_1) \\ k_4 \cdot S_2 \end{pmatrix}$$

C regulated production

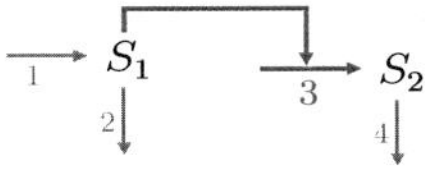

$$\frac{dS_1}{dt} = k_1 - k_2 \cdot S_1$$
$$\frac{dS_2}{dt} = \boxed{k_3 \cdot f(S_1)} - k_4 \cdot S_2$$

$$\Leftrightarrow \quad \begin{pmatrix} \frac{dS_1}{dt} \\ \frac{dS_2}{dt} \end{pmatrix} = \begin{pmatrix} 1 & -1 & \boxed{0} & 0 \\ 0 & 0 & \boxed{1} & -1 \end{pmatrix} \cdot \begin{pmatrix} k_1 \\ k_2 \cdot S_1 \\ k_3 \cdot f(S_1) \\ k_4 \cdot S_2 \end{pmatrix}$$

Figure 5.1.: Comparison between pure production, conversion and regulated production for an example system with two variables S_1 and S_2. The reaction marked in red with index 3 is either a pure production (A), a conversion mediating matter flow from S_1 to S_2 (B) or a regulated production which does not mediate matter flow between S_1 and S_2 (C). Left: Visualizations of the ODE systems in the center. The red boxes in the ODE systems mark the rate functions of reaction 3. f denotes a non-zero function depending positively on S_1. Right: Matrix representation of the ODE systems with stoichiometric matrix. The red boxes mark therein the entries for reaction 3.

ical process alters at least one of the concentrations of the species considered in the model, otherwise the reaction or process would be neglected in the model. Therefore, by definition, any reaction in a mathematical model mediates a certain matter flow. On the one hand, there are processes altering only the number of molecules or concentration of exactly one species. Typical such processes are degradation reactions which only decrease the concentration of the degraded species (reactions 2 and 4 in Figure 5.1), and pure production reactions which only increase the concentration of the produced species while not affecting any other species of the model (reaction 3 in Figure 5.1 A and reaction 1 in Figure 5.1 A-C). On the other hand, there are processes altering the concentration or molecule number of more than one species in the model or biological system. Rarely, a reaction representing such a process only either decreases or produces multiple species. Instead, such a reaction typically decreases the concentration of some of the species (the source species) while increasing some of the concentrations of other species (the product species). In this case a conversion of matter between at least two species is taking place, the source species of the reaction are consumed when being converted to the product species. Such a reaction is called a *conversion reaction*, or simply *conversion* (for example reaction 3 in Figure 5.1 B). Metabolic processes are examples for conversion reactions, as well as transport processes if the species in the according compartments are distinguished, and modifications like phosphorylations, methylations, ubiquitinylations in which the unmodified species concentration is decreased while the concentration of the modified species increases. A conversion reaction acts like a degradation reaction on the source species and like a production reaction on the product species, both with exactly the same rate (up to stoichiometry) and both being simultaneously influenced if the conversion reaction is perturbed.

Regarding its wiring and connection in the model or biological system, a conversion reaction resembles a production where one or more species positively regulate the otherwise pure production, a *regulated production* (reaction 3 in Figure 5.1 C). Similar to a regulated production, the production of the product species of the conversion is depending positively on its source species, it is positively regulated by them. The only difference between a regulated production and a conversion is that in the first, the regulating species are not consumed via the reaction, the matter flow from the regulating species to the product species is disrupted. In fact, the ODE systems representing a conversion reaction or a regulated production reaction can be very similar (compare Figure 5.1 B and C). The one can replace the other and thus the matter flow properties of the reaction in the system can be changed by simply altering the according entries of the reaction in the stoichiometric matrix for the source species or regulating species (compare matrix representation of the ODE systems in Figure 5.1 B, C and section 2.1.1). If a reaction is a conversion, the regulating species are source species of the reaction and thus are reduced by the reaction which is indicated by a negative entry in the stoichiometric matrix. For the equations of the ODE system this implies that the term for the reaction rate occurs not only in the equation of the product species, but with a minus sign also in the equation of the source species (Figure 5.1 B). If the reaction is a regulated production and hence matter flow is lacking from the regulating species to the product species, the regulating species are not directly influenced by the reaction, the according entries in the stoichiometric matrix are zero. This means that the reaction rate term only occurs in the equation of the product species even though the reaction rate depends on other species (Figure 5.1 C).

Thus, with merely minor changes in the stoichiometric matrix, a reaction in a model can be altered from a regulated production to a conversion and the other way round. This means that for these cases, the matter flow property of the reaction can be easily switched from lacking matter flow between species to mediating matter flow between species and *vice versa*. Note that only a *positively* regulated production is this easily changed into a conversion in terms of the ODE systems. As soon as the regulation of the production by a species is negative, the regulating species cannot be a source species of this reaction in biological models since for plausibility reasons, all reactions depend positively on all of their source species (compare section 2.1.3). Additionally, in contrast to switching from a regulated production to a conversion to alter the matter flow properties, for changing a pure production which as well by definition lacks matter flow between species into a conversion or *vice versa*, also alterations in the defining functions for the reaction rates are necessary (compare Figure 5.1 A, B). For these reasons, this work restricts the analysis of the effects of matter flow between species to the differences arising when employing conversions opposed to the according positively regulated productions.

5.1.1. Matter flow vs. flow of information in biology

In fact, the whole issue of lacking matter flow between species is due to simplification which is necessary for the examination of biological systems. If considering each and every biological entity separately, every single process mediates a matter flow between two entities – this is a physical law, matter cannot just vanish, nor be created from literally nothing. For example, the degradation of a protein is performed step-wise by the cellular proteasome, breaking the amino acid chain into smaller pieces until only single amino acids remain which can be and are used again for the synthesis of other (or the same) proteins. Just the simplification of the system by neglecting smaller amino acid chains and single amino acids results in the degradation reaction lacking matter flow between species.

In different biological systems, different simplifications are made, and conversion reactions and regulated productions represent two opposing biological mechanisms:

Matter flow versus pure flow of information (in the following referred to only as flow of information) - matter flows from upstream species to product species versus the production of a downstream species is only regulated by some upstream species. Thereby, matter flow from a source species to a product species is generally assigned to metabolic systems whereas flow of information is supposed to prevail in signaling (Ihekwaba et al. [2004], Klipp and Liebermeister [2006]). This is due to the nature of the examined systems. Metabolism constitutes by definition chemical transformations, the conversion from metabolites to other metabolites, e.g. from glucose to shorter carbohydrates as pyruvate in glycolysis (Higgins [1964], Ghosh and Chance [1964]). Therefore, simplifications resulting in regulated productions make no sense because explicitly the conversions from one metabolite to the other are of central interest. In signal transduction, however, it is more focused on how information is transferred. A signaling molecule outside the cell may be subject to endocytosis as the epidermal growth factor together with its receptor (Sorkin and von Zastrow [2002]) and may be further transported to other compartments of the cell. Both implies that the matter of the signaling molecule flows. However, the response to a signal will most probably not be only the chemical transformation of the signaling molecule to some other species and its subsequent exocytosis. Instead, the signaling molecule could act as transcription factor thereby regulating the expression of target genes, i.e. regulating the production of one or more proteins. At least at this point, simplifications are routinely made which gather the transcriptional (and sometimes also translational) machinery into one production of mRNA (or protein) regulated by the transcription factors. Thus, matter flow from the signaling molecule to the product species constituting the readout of the response to the signal is disrupted. Even more frequently in signaling, already at earlier stages of signal transduction simplifications are implemented and regulated productions instead of conversions occur. For example, the signaling molecule may not be transported into the cell but it is frequently assumed to only affect a receptor on the cell membrane as in the non-canonical NF-κB stimulus altering the lymphotoxin β receptor (Sun [2011]). Still, in signal transduction and metabolism both types, conversions and regulated productions, can occur.

The exemplary oscillatory systems introduced in chapter 3 which crucially differ in their period sensitivities also differ in the matter flow properties of their basic mechanisms. Calcium oscillations are mainly resulting from transport reactions which mediate matter flow between two species, e.g. calcium from the ER is transported to the cytosol (and back) thereby reducing the concentration of endoplasmic calcium and increasing the concentration of cytosolic calcium. In contrast, as circadian oscillations rely on a transcription-translation feedback loop (compare section 3.1.1) whose details concerning transcription and translation are commonly neglected, matter flow between the species in the system, but not information flow, is disrupted. For example, the clock protein is not the product species of a conversion, but of a regulated production regulated by the clock protein mRNA which is not consumed via the process of translation represented in one single reaction. In the following, with the help of the chain models introduced in section 4.2, the effect of the occurrence or lack of matter flow between species on the sensitivity of the period and the amplitude in oscillating systems is investigated.

5.2. Matter flow in the chain models

In the chain models introduced in section 4.2 and given in Figure 4.1, different reactions are considered. The constant production reaction ν_1 only mediates matter flow to species S_1, while the four degradation reactions $\nu_3, \nu_5, \nu_7, \nu_8$ only mediate matter flow from the according degraded species. In the already examined version of the chain model (section 4.2.1), there are three conversion reactions, ν_2, ν_4, and ν_6, in which species S_1, S_2 and S_3 are transformed to their subsequent species in the chain. Without applying major changes as introducing further species and/or reactions and parameters to the chain models, the matter flow properties of only these three reactions, ν_2, ν_4, ν_6, can be altered.

The step-wise variation of the matter flow properties of reactions ν_2, ν_4, and ν_6 from conversion reactions as in the original chain models to regulated productions of species S_2, S_3 and S_4, respectively, regulated by species S_1, S_2 and S_3, respectively, results in eight chain models for each feedback type (Figure 5.2). The models are named according to whether reaction ν_i includes matter flow between species and is a conversion (index "i") or whether it lacks matter flow between species ("-"). For example, C246 codes for the original chain models with all three reactions being

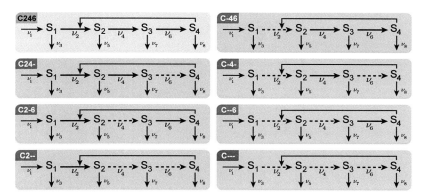

Figure 5.2.: Model structures for the chain models with different matter flow properties. Each model is labeled according to which reactions of ν_2, ν_4 and ν_6 are conversion reactions and hence mediate matter flow between species. For simplification, regulated productions are indicated by dashed arrows from the regulating species to the product species. The model structure C246 colored in orange is that of the original models defined in chapter 4.

conversions, C2-6 for the models with ν_4 being a regulated production (and hence lacking matter flow between species) and C--- for the models with all three reactions ν_2, ν_4, ν_6 being regulated productions instead of conversions. The according ODE systems are given in the Appendix (section A.7.2).

5.2.1. Matter flow properties influence the regulatory structure

First it is asked whether all eight chain models from Figure 5.2 for both negative or positive feedback can exhibit sustained oscillations. As described in section 5.1, alterations of the matter flow properties of reactions necessitate only minor changes of the stoichiometric matrix of the ODE system. However, these minor changes can result in decisive differences in the regulatory structure of the system, especially if the matter flow property of a reaction is altered which is subject to additional regulatory interactions. For the chain models, this concerns reaction ν_2 on which S_4 acts negatively or positively for the negative feedback or positive feedback chain model, respectively. When comparing the Jacobian matrices of models C246 and C---

(Appendix, Figure A.10), it becomes evident that for ν_2 lacking matter flow between species, the feedback loop comprising all four species is abolished in the chain model with either type of feedback (compare section 4.2). For the negative feedback chain models, this means that the only feedback loop present in the system is the negative feedback loop comprising species S_2, S_3 and S_4, and thus the system still permits the occurrence of sustained oscillations. In contrast, if ν_2 lacks matter flow between species, the positive feedback chain models yield only a positive feedback loop (via species S_2, S_3, S_4) but no negative feedback loop thereby precluding the existence of sustained oscillations (compare section 4.1). Therefore, for the positive feedback chain models, only models C246, C24-, C2-6 and C2-- (on the left in Figure 5.2) need to be considered here since these are the only ones in which sustained oscillations may occur.

5.2.2. Is a further simplification of the analysis possible?

The ODE systems of the chain models with different matter flow properties are very similar to each other (compare section 5.1 and the ODE systems for the chain models given in the Appendix, section A.7.2). Indeed, any biologically feasible solution (i.e. positive solution for positive parameter values) which can be obtained in one of the chain models from Figure 5.2 can also be obtained in model C2-- or model C--- (elaborated in the Appendix, section A.3.1). Thus, one could be tempted to conclude that also for the sensitivity analysis, an examination of these two models for each feedback type would suffice. However, this conclusion is not valid as shown for the negative feedback chain models C246 and C2-- (Appendix, section A.3.2). At the chosen example parameter sets, both models yield exactly the same solution and hence exactly the same limit cycle oscillations, but the sensitivities are different for the different model structures. This difference is due to the fact that the effect of an applied parameter perturbation depends on the particular encoding of the perturbed reaction, and this can differ in two ODE systems yielding originally the same solution. Therefore, it is not sufficient to analyze the two chain models C2-- and C--- for each feedback type which yield all feasible solutions, but all eight (for negative feedback) or four (for positive feedback) combinations of matter flow properties in reaction ν_2, ν_4 and ν_6 have to be considered in order to exhaustively determine the effect of matter flow properties on the period sensitivity and amplitude sensitivity in the chain models.

5.3. Sensitivity analysis of the chain models with different matter flow properties

For the sake of clarity, for each feedback type, the analysis in the main text is only given for the comparison of the two oscillating chain models which yield the utmost number of reactions differing in their matter flow properties (compare section 5.2.1). Hence, only the results of the sensitivity analysis of model C--- with negative feedback and model C2-- with positive feedback are shown in comparison to the results for model C246 with the according feedback type (Figure 5.3). The analysis for variations of the matter flow properties of single reactions can be found in the Appendix (sections A.3.3 and A.3.4).

The period and amplitude sensitivities obtained for the negative feedback chain model C246 (orange stars) compared to model C--- (red squares) are given in Figure 5.3 A together with the according box-plots of the sensitivity distributions, and in the Appendix, Tables A.10, A.12. If disrupting matter flow between species in all three reactions, the median period sensitivity increases by 5% from 0.189 to 0.198 (Table A.10). Although being only slight, the increase in the period sensitivity is significant (MWU p-value: $1.5 \cdot 10^{-259}$, Table A.12). The dispersion of the period sensitivity distribution measured by the 90%R reduces by three-fourths from 0.05 for model C246 to 0.012 for model C--- (Table A.10). The median amplitude sensitivity rises only slightly by 8% from 0.66 for model C246 to 0.71 for model C---. The increase in the amplitude sensitivity distribution is significant (MWU p-value: $3.6 \cdot 10^{-6}$, Table A.12). The dispersion for the amplitude sensitivity distribution is nearly the same for both models, the 90%R decreases only by 4% from 1.49 for model C246 to 1.43 for model C--- (Table A.10), although visibly lower maximal and larger minimal amplitude sensitivity values are reached for the model lacking matter flow between species (box-plot to the right in Figure 5.3 A). The correlation between period sensitivity and amplitude sensitivity for the negative feedback chain model is strengthened decisively from $R_S = 0.30$ for model C246 to $R_S = 0.94$ for model

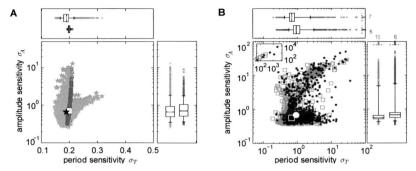

Figure 5.3.: Impact of matter flow on the sensitivity of the chain models. A: Sensitivities for the negative feedback chain model C246 (orange stars, black star as median values) or model C--- (red squares, white square as median values). The box-plots in the according colors summarize the characteristics of the distributions of the period sensitivity (top) and amplitude sensitivity (right). B: Sensitivities for the positive feedback chain model C246 (green squares, black square as median values) or model C2-- (dark green dots, white circle as median values), as well as the respective box-plots.

C--- indicating a strong correlation. The correlation is larger the more reactions lack matter flow between species (Table A.10).

The period and amplitude sensitivities obtained for the positive feedback chain models C246 (green squares) and C2-- (dark green dots) are shown in Figure 5.3 B together with the according box-plots of the sensitivity distributions, and in the Appendix, Tables A.11, A.13. If lacking matter flow between species in ν_4 and ν_6, the median period sensitivity rises by nearly 40% from 0.68 for model C246 to 0.93 for model C2-- (Table A.11). The increase observed in the period sensitivity distributions is significant (MWU p-value: $3.3 \cdot 10^{-180}$, Table A.13). The 90%R of the period sensitivity distribution increases by one-fourth from 1.99 for model C246 to 2.50 for model C2-- (Table A.11). Concerning the amplitude, the median amplitude sensitivity mounts by nearly one-fifth from 0.57 for model C246 to 0.68 for model C2-- (Table A.11). The increase in the amplitude sensitivity distribution is significant (MWU p-value: $7.5 \cdot 10^{-89}$, Table A.13). The 90%R for the amplitude sensitivity distribution rises by 16% from 4.48 for model C246 to 5.20 for model C2--

(Table A.11). The correlation between period sensitivity and amplitude sensitivity for the positive feedback chain model is weakened yielding a value of $R_S = 0.42$ for model C246 and $R_S = 0.14$ for model C2--. The correlation is smaller the more reactions lack matter flow between species (Table A.11).

In summary, introducing reactions which lack matter flow between species increases both period and amplitude sensitivity for the chain model with either feedback type. Thereby, the median of the period sensitivity distribution of the negative feedback chain model is affected least; its dispersion is affected (decreased) most. In the next section, the origins of the increased sensitivities in the chain model with either feedback type and the strongly altered dispersion of the period in the negative feedback chain model are investigated.

5.4. Origins of particular effects of matter flow on sensitivity

5.4.1. Increase of sensitivities in the negative feedback chain models

In the negative feedback chain models, replacing conversions by regulated productions slightly increases both period sensitivities and amplitude sensitivities (Figure 5.3 A). Considering the contribution of every individual parameter on the overall sensitivity by examining their sensitivity coefficients reveals that the sources of the slight increases in the sensitivities are different for period sensitivity and amplitude sensitivity (Figure 5.4).

In the model with regulated productions instead of conversions (model C---) the median period sensitivity coefficients of rate coefficients 5, 7 and 8 are large compared to those of all other parameters (Figure 5.4 A, right). This indicates that for the negative feedback chain model, the period is almost exclusively affected by the rate coefficients of the reactions mediating a decrease in species S_2, S_3 or S_4. In model C246, these are not only the velocities in degradation reactions ν_5, ν_7 and ν_8 as in model C---, but additionally rate coefficients 4 and 6 because the according reactions ν_4 and ν_6 are conversions and mediate by definition a decrease in their source species (compare model scheme in Figure 5.4 C). Accordingly, for model C246, rate coefficients 4 - 8 contribute considerably to the period sensitivity (Figure 5.4 A, left). However, if summing the absolute period sensitivity coefficients over the parameters $(\sum_{i=1}^{8} |R_i^T| + |R_{kn}^T|)$ for each parameter set, even slightly lower values are reached for model C--- than for model C246 (see Figure 5.4 D, $\mu(\sum_{i=1}^{8} |R_i^T| + |R_{kn}^T|) =$ 1.00 for model C--- opposed to 1.02 for model C246, MWU p-value $3 \cdot 10^{-262}$). Hence, the increase in period sensitivities obtained for the model lacking matter flow between species is not caused by increasing sensitivities for changes in the individual parameters, but merely by a redistribution of the sensitivities among the parameters. Since calculation of the overall sensitivity σ_T involves squaring the period sensitivity coefficients (see the defining Equation 2.6) and since for all non-negative real values a, b the square of the sum is greater than or equal to the of sum of the squares $((a + b)^2 = a^2 + 2ab + b^2 \geq a^2 + b^2)$, the value of the sensitivity σ_T depends not only on the sum of the absolute sensitivity coefficients, but also on the distribution of the sensitivity among the parameters. The less uniformly the sensitivity is distributed, i.e. the less parameters contribute to the sensitivity (while yielding the same sum of absolute sensitivity coefficients), the greater is the sensitivity. Model C--- has a lower number of parameters mediating a decrease in species S_2, S_3, S_4 and thus contributing to the period sensitivity than model C246. Consequently, even though the sum of $|R^T|$s is slightly smaller for model C--- than for model C246, the σ_Ts of model C--- are slightly larger than those of model C246.

Regarding the amplitude sensitivities of the negative feedback chain models, their increase for lacking matter flow between species is mainly emerging from an actual increase in the sensitivities of the amplitude towards changes in particular parameters (Figure 5.4 B). The amplitude sensitivities are again redistributed among the parameters by introducing reactions lacking matter flow: While in model C246 rate coefficients 4, 6, 8 and nl-parameter kn_1 display the highest median $|R^A|$ (Figure 5.4 B, left), rate coefficients 5, 7, 8 and also kn_1 dominate the amplitude sensitivities in model C--- (Figure 5.4 B, right). However, in model C246 also rate coefficients 5 and 7 and in model C--- also rate coefficients 4 and 6 contribute to the amplitude sensitivities. Thus, the number of parameters considerably contributing to the amplitude sensitivities comparing the two models is not as impaired as for the period sensitivities if introducing lack of matter flow between species. Consequently, the observed increase in amplitude sensitivities can be mainly attributed to an actual increase in the amplitude sensitivities towards perturbations of the single parameters. Summing over the absolute values of the amplitude sensitivity coefficients delivers slightly but significantly larger values for model C--- compared to model C246 (see Figure 5.4 D, $\mu(\sum_{i=1}^{8} |R_i^A| + |R_{kn}^A|) = 4.56$ for model C--- opposed to 4.36 for model C246, MWU p-value $3.7 \cdot 10^{-4}$). Still, there is possibly also a small effect of the redistribution of sensitivities included as the difference in median values comparing the sums $\sum_{i=1}^{8} |R_i^A| + |R_{kn}^A|$ is only 4.7% while for the σ_A, the difference is 8%.

Comparing the models with different matter flow properties, also a change of the sensitivity coefficients of rate coefficients $1 - 3$, i.e. parameters $k_1 - k_3$, can be observed. In contrast to model C246, in model C---, the rate coefficients $1 - 3$ yield always the same absolute sensitivity coefficients (right panels in Figure 5.4 A, B). The reason for this is that S_1 is decoupled from the only negative feedback loop in the system consisting of S_2, S_3 and S_4 in model C---, i.e. S_1 is not affected by any of the other species (compare the ODE for S_1 for model C--- in the Appendix,

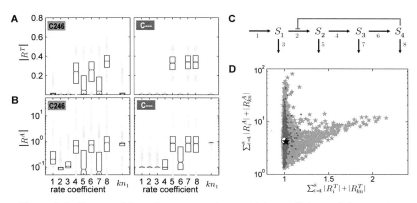

Figure 5.4.: Impact of flow of matter on the sensitivity coefficients of the negative feedback chain model. A, B: Box plots of the absolute period and amplitude sensitivity coefficients, respectively, for models C246 (left) and C--- (right). C: Scheme of model C246, indices of reactions accord with indices of rate coefficients. D: Sum of the absolute sensitivity coefficients over the eight rate coefficients and the nl-parameter kn_1, $\sum_{i=1}^{8} |R_i| + |R_{kn}|$, for each parameter set of model C246 (orange stars, median as black star) compared to model C--- (red dots, white circle for median values).

section A.7.3). Accordingly, S_1 is not oscillating together with the other species in the system but remains constant after a transient phase, $S_1 = k_1/k_3 = const.$ Recall that $\nu_2 = k_2 \cdot fb \cdot S_1$, with fb being the feedback term. Hence, perturbations in k_2 have exactly the same effect as similar perturbation in k_1 on the system, and consequently, also the sensitivity coefficients of k_1 and k_2 coincide. Furthermore, increasing k_3 by 2% results in a decrease of S_1 by 1.96% which results in almost the same perturbation strength as if altering k_1 or k_2 by 2%, only in the opposite direction. This makes no difference for the sensitivity coefficients for small perturbations as applied here (compare sections 2.5.4, A.1.1), and on that account, the absolute sensitivity coefficients of k_3 are similar to those of k_1 and k_2.

5.4.2. Decrease in dispersion of the period sensitivity in the negative feedback chain models

The strongest influence of introducing reactions lacking matter flow on the variability of the sensitivities can be observed for the period sensitivity distribution of the negative feedback chain models where the 90%R is reduced by 75%. In fact, the variability is reduced to both sides of the median. Neither parameter sets with period sensitivities much higher nor parameter sets with period sensitivities much lower than the median occur in model C--- opposed to model C246 leading to an extremely small variability of the period sensitivity in model C--- (compare Figure 5.3 A).

In the following, the origins of this reduction in period sensitivity variability in the negative feedback chain models are investigated. Consulting the results of the sensitivity analyses for the negative feedback chain models with only one or two conversions being replaced by regulated productions (Appendix, Figure A.13), it becomes obvious that the variability concerning period sensitivities greater than the median is influenced mainly by the matter flow property of reaction ν_2 while the variability concerning period sensitivities smaller than the median is mainly affected by the matter flow properties of reactions ν_4 and ν_6.

The influence of the matter flow properties in ν_2 is evident if recapitulating that all parameter sets in the negative feedback model C246 yielding period sensitivity values considerably higher than the median originate from a population of parameter sets with $max(\nu_2) > max(\nu_3)$ (section 4.4.1). In section 4.4.1, it has been discussed that these parameter sets display higher period sensitivities since S_1 is rather processed through the chain instead of degraded via ν_3. Thereby, the sensitivities of the period (and amplitude) towards particular parameters are increased. As soon as ν_2 is lacking matter flow between species, S_1 can exclusively be diminished via ν_3 and it is never processed through the chain. Thus, also the increase in period sensitivities associated with large ν_2 compared to ν_3 cannot occur in any of the models lacking matter flow in ν_2, and consequently also not in model C---.

Concerning the influence of the matter flow properties in reactions ν_4 and ν_6, a

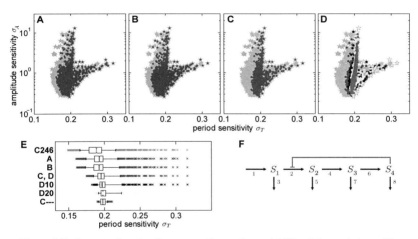

Figure 5.5.: Impact of matter flow properties on the variability of the period sensitivity of the negative feedback chain models. A-D: Sensitivities for model C246 (orange stars). A, B: Sensitivities for those parameter sets of model C246 with $\nu_4 \gg \nu_5$ or $\nu_5 \gg \nu_4$ (A, 1663 parameter sets), $\nu_6 \gg \nu_7$ or $\nu_7 \gg \nu_6$ (B, 1754 parameter sets) as dark red dots. $\nu_i \gg \nu_j$ means $\nu_i > a \cdot \nu_j$ with $a = 5$. C, D: Sensitivities for the parameter sets with either $\nu_4 \gg \nu_5$ or $\nu_5 \gg \nu_4$ and either $\nu_6 \gg \nu_7$ or $\nu_7 \gg \nu_6$ (dark red dots, 1187 parameter sets). Black and gray dots in D indicate sensitivities for the parameter sets similar to C, but obtained from model C246 by using $a = 10$ (black, labeled D10 in E, 868 parameter sets) or $a = 20$ (gray, labeled D20 in E, 601 parameter sets) in the definition of "$\gg$". Red dots indicate the sensitivities for model C---. E: Box-plots of the period sensitivities obtained in A-D. F: Model scheme of the negative feedback chain model C246.

result from the previous section 5.4.1 is used. Therein, it has been observed that the period sensitivities are nearly exclusively determined by the reactions mediating a decrease in species S_2, S_3 and S_4. In particular, the differences in period sensitivities between models C246 and C--- arise solely due to a redistribution of the sensitivities among the parameters (Figure 5.4). The less parameters considerably contribute to the period sensitivities, the higher the period sensitivities. These findings also explain the decrease in the variability concerning period sensitivities with values lower than the median. The difference between model C246 and C--- is the number of reactions mediating a decrease in species S_2 or S_3, which is ν_4 and ν_5 or ν_6 and ν_7, respectively, in model C246, but only ν_5 or ν_7 in model C---. Accordingly, Figure 5.5 displays how the period sensitivity in model C246 is affected by the relation between ν_4 and ν_5 or ν_6 and ν_7. If the values of ν_4 and ν_5 are very distinct, S_2 is either diminished nearly exclusively via ν_4 or exclusively via ν_5. This results in an uneven distribution of the period sensitivities between the two parameters. The $|R^T|$s of the parameter mainly governing the decrease in S_2 are large and those of the other are small (not shown). Consequently, the period sensitivities and thus also the lower boundary for the period sensitivities are increased (Figure 5.5 A, E). The same applies for the relation between ν_6 and ν_7 (Figure 5.5 B, E). The lower boundary of the observed period sensitivities is even further shifted to higher values if both pairs yield unevenly distributed $|R^T|$s (Figure 5.5 C, E). The additive effects in the diminishing reactions of S_2 and S_3 are similarly observed for the models with only one or two reactions lacking matter flow (Appendix, Figure A.13). Model C--- is the extreme case of the period being only sensitive to k_5 and k_7 but not to k_4 or k_6. The variability concerning period sensitivities lower than the median approaches that of model C--- if the reaction rates in the pairs ν_4, ν_5 and ν_6, ν_7 differ more strongly (Figure 5.5 D, E).

5.4.3. Increase of sensitivities in the positive feedback chain models

Lacking matter flow between species leads to an increase in both period sensitivities as well as amplitude sensitivities of the positive feedback chain models (Figure 5.3 B). Indeed, this effect is already visible on the level of the sensitivity coefficients of

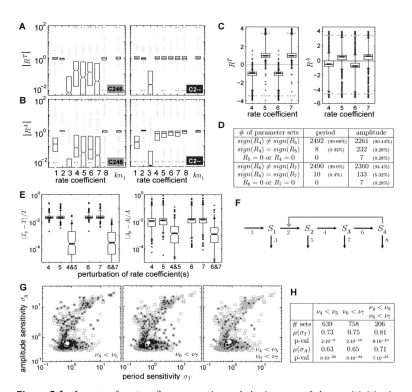

Figure 5.6.: Impact of matter flow properties and the increase of the sensitivities in the positive feedback chain models. A, B: $|R^T|$s (A) and $|R^A|$s (B) for the positive feedback chain models C246 (left) and C2-- (right). C: R^Ts and R^As for model C2--. D: Number of parameter sets (and percentages) of model C2-- for which perturbations in k_4 and k_5 or k_6 and k_7 yield opposite changes in the period or amplitude or not. E: Relative changes (perturbed value T_p, A_p minus original value T, A) in period or amplitude for increasing k_4, k_5, or both (4&5), k_6, k_7, or both (6&7) by 2% for 250 random parameter sets of model C2--. F: Model scheme of the positive feedback chain model C246. G: Sensitivities of model C246 for those parameter sets with $\nu_4 < \nu_5$, $\nu_6 < \nu_7$ or both (dark green dots, white circle as median values). Light green squares, black square for median values: Sensitivities for all 2500 parameter sets of model C246. H: Median sensitivities for subsets of model C246 according to panel F and p-values for the comparison to the sensitivities obtained for all 2500 sets.

the individual parameters which are given for models C246 and C2-- in Figure 5.6 A and B. For both oscillatory characteristics, the absolute sensitivity coefficients of rate coefficients 4 - 7 are decisively increased in model C2-- (right panels) compared to model C246 (left panels). Thus, in contrast to the negative feedback chain models where reactions diminishing species concentrations are affecting the period and amplitude most strongly, for the positive feedback chain models also productions are critical for the sensitivities. The striking difference in the sensitivities especially towards rate coefficients 4 and 6 comparing the model with and without matter flow between species emerges from a lack of compensation. This compensation is present if altering a conversion which mediates a production in combination with a decrease in the preceding species in the chain, but it is lost if exclusively altering a regulated production. This is revealed by the sensitivity coefficients of model C2-- for parameters k_4 - k_7 (Figure 5.6 C). Increasing parameters k_4 or k_6 leads for more than 99% of the parameter sets to an opposite change in the period than increasing parameters k_5 or k_7 (Figure 5.6 D). For the amplitude, this is the case for more than 90% of the parameter sets. Accordingly, a compensatory effect is restored for model C2-- if perturbing k_4 and k_5 or k_6 and k_7 simultaneously in the same direction. In these cases, the relative changes evoked in period or amplitude are three orders (period) or one order (amplitude) of magnitude lower compared to perturbing the parameters separately (Figure 5.6 E). The compensatory effect can also be reduced in model C246 if the conversion rate is smaller than the according degradation rate, i.e. if $\nu_4 < \nu_5$ or $\nu_6 < \nu_7$. If the conversion ν_4 or ν_6 is perturbed under these conditions, the whole production rate of S_3 (ν_4) or S_4 (ν_6) is perturbed, but only less than half of all reactions diminishing S_2 ($\nu_4 + \nu_5$) or S_3 ($\nu_6 + \nu_7$) is perturbed simultaneously in the same direction and thus compensation is weak. For this reason, both period and amplitude sensitivities are significantly increased for $\nu_4 < \nu_5$ or $\nu_6 < \nu_7$ in model C246, and even more if both conditions are met simultaneously (Figure 5.6 G and H, MWU p-values $< 10^{-9}$ for the period sensitivity, $< 10^{-25}$ for the amplitude sensitivity).

Thus, the increase in period and amplitude sensitivities for the positive feedback model lacking matter flow between species can be attributed to an emerging lack of compensation of producing and decreasing processes. Both types of processes are mediated by conversions and thus both processes are at least in part simultaneously perturbed in model C246 if applying a single parameter change, whereas in model C2-- producing and decreasing processes are more stringently separated and in general, only one of the processes is altered if perturbing a single parameter. Since the sensitivity calculation encompasses single parameter perturbations only, the model lacking matter flow between species displays higher sensitivities.

5.5. Summary and discussion: The effect of matter flow on sensitivity

In this chapter, the effects on period and amplitude sensitivities are examined which occur for employing reactions lacking matter flow between species, i.e. regulated productions, instead of reactions mediating a matter flow between species, i.e. conversions. First of all, altering the matter flow properties in such a way can change the regulatory structure of the underlying ODE system, especially if the conversions are additionally regulated (section 5.2.1). Matter flow between species then potentially has an impact on whether feedback loops exist or not. Conversions can increase the number of feedback loops present. Consequently, the presence of matter flow between species can be decisive for the occurrence of sustained oscillations. Indeed, matter flow between species is crucial for systems with positive feedback, and otherwise positive interactions only, to create a negative feedback loop which is necessary for the system to yield sustained oscillatory dynamics. In particular, for the positive feedback chain model, substituting the conversion ν_2 by a regulated production leads to the lack of limit cycle oscillations (section 5.2.1).

Altering the matter flow property of reactions also affects the period sensitivities and amplitude sensitivities as shown in the chain models. Indeed, irrespective of the feedback type, lacking matter flow between species promotes the occurrence of larger period and amplitude sensitivities (section 5.3). The sources of these increases which are partly only slight are different depending especially on the particular model structure.

For the negative feedback chain models, the analysis of model C--- lacking matter flow between species reveals that both the period and the amplitude in this model are most sensitive to processes mediating the decrease or decay of the inner-loop species

S_2, S_3 and S_4 (section 5.4.1). Lacking matter flow between species results in fewer parameters mediating the decrease of these species while the sum of the sensitivities remains almost constant. The sensitivities are only redistributed among the parameters. The calculation of the overall sensitivities σ_T and σ_A by the quadratic mean results in slightly increased period sensitivities and increased amplitude sensitivities for these more asymmetric distributions of sensitivity coefficient values among the parameters of models lacking matter flow compared to model C246 (section 5.4.1, at most 6% and 19% increase, respectively, for any possible variation in matter flow properties in ν_2, ν_4, ν_6, Table A.10). Additionally, for the negative feedback chain models, the dispersion of the period sensitivity distributions is decisively reduced in models employing regulated productions instead of conversions. Period sensitivities higher than the median are less frequently obtained if ν_2 is a regulated production instead of a conversion. Period sensitivities lower than the median occur more rarely as soon as ν_4 and/or ν_6 lack matter flow between species because of similar redistribution effects as described above (section 5.4.2).

For the positive feedback chain model C2--, it is observed that the period as well as amplitude are sensitive to perturbations which affect only a decrease or only a production of a species (section 5.4.3). Thereby, perturbations of degradations are found to most frequently generate changes of the period or amplitude in the opposite direction than perturbations of the productions of the subsequent species in the chain. Consequently, a compensatory effect is observed: perturbations of conversions which mediate by definition both, production and at least a part of the decrease of the preceding species in the chain, alter the oscillatory properties decisively less than perturbations of reactions which only decrease or only produce species (section 5.4.3). As a result, the sensitivities of the chain models lacking matter flow between species and thus also lacking the compensatory effect are drastically increased compared to model C246 (up to 36.8% for the period and 19.3% for the amplitude sensitivities for any possible variation in matter flow properties in ν_2, ν_4, ν_6, Table A.10).

Taken together, differences in matter flow properties change the sensitivities and even more decisively the sensitivity coefficients obtained for the chain models. Thus, it can be assumed that matter flow properties are also important for the sensitivities of the period and of the amplitude of other models.

5.5.1. Matter flow from more than one source species

In biological models, a species is a potential source species of a reaction if the reaction rate depends positively on the species (compare section 2.1.3). In the chain models, due to their simplicity, a reaction can potentially yield only one source species (up to reaction ν_2 in the positive feedback chain model). Therefore, regarding the matter flow properties of a reaction, only two options are possible in the chain models: The reaction lacks matter flow between species, or it includes matter flow with the source species being the only species possible. If the number of potential source species of a reaction in a model is larger than one, there is more than one option for the scenario of the reaction to mediate matter flow between species. For example, if a reaction depends positively on species S_1 and S_2 (e.g. reaction ν_3 in the example system in Figure 2.1, section 2.1.3), it includes matter flow between species if (i) S_1 is a source species or (ii) S_2 is a source species or (iii) S_1 and S_2 are source species of the reaction. The effect of the choice of the particular option on sensitivity is not examined in the scope of this thesis. However, differences in sensitivities between different models in which the reaction in question mediates matter flow between species but which employ option (i), (ii) or (iii) are to be expected. For each option, the contribution of kinetic parameters is distributed differently among the processes modeled in the system enabling more or less compensation of single parameter perturbations. Of course, processes are least decoupled if applying option (iii) and hence, having the results from the examination of the chain model in mind, most compensatory effects and therefore possibly most reduced sensitivities are expected for the reaction mediating matter flow from both species to the product species compared to the model lacking matter flow between species. However, the actual effect of such differences in the assumptions for the matter flow properties of reactions needs to be evaluated in the context of the examined model.

In principle, the positive feedback chain model delivers a possibility to examine the influence of having two source species and one product species. In the model, reaction ν_2 where the feedback applies is a conversion from S_1 to S_2 which is positively regulated by S_4. Therefore, it is easily possible to establish matter flow also from S_4 to S_2 via ν_2. However, this positive feedback chain model does not oscillate (data not shown).

5.5.2. Lacking matter flow in signaling processes

For models of metabolic systems, matter flow between species is usually considered to occur in any reaction. However, in models of signaling systems, closed loops, where matter flows, are combined with reactions carrying a flow of information but lacking matter flow (Klipp and Liebermeister [2006]). In this chapter, it has been figured out that lacking matter flow can lead to increased sensitivities. This happens either by redistribution of the sensitivities among the parameters as in the negative feedback chain model, or by really increasing the sensitivity of the system through decoupling of processes and preventing compensation. Therefore, considering the results of this section, one should be aware that by introducing reactions lacking matter flow in signaling processes for simplification, the sensitivity properties of the examined process might be changed and reveal possibly larger sensitivities than may be observed in nature.

5.5.3. Applicability of the results

In this work, reactions mediating a matter flow between species have been replaced one to one by reactions lacking matter flow between species. For a biological process, however, this would not be feasible since reactions lacking matter flow always sum up or simplify several intermediate reaction steps. These simplifications are usual practice in systems biology, for example for transcription and translation. They are acceptable if the single steps of a process are considered not to affect the behavior which is to be examined in the model, or if details or information about the steps in the process are not available. Therefore, the approach of replacing a reaction including matter flow one to one by a reaction lacking matter flow applied on the chain models in this work is, strictly biologically seen, not valid. However, it is useful to unravel the effects the introduction of reactions lacking matter flow can exert. Perspectively, it would be interesting to examine the differences occurring in the sensitivities of the period and amplitude if comparing a more detailed mechanism yielding reactions only including matter flow to the simplification of these details by one reaction lacking matter flow.

6. The effect of saturating kinetics on sensitivity

In the preceding chapters 4 and 5, structural principles as the type of feedback and matter flow and their effect on the sensitivity of the period and amplitude of oscillations have been examined. In this chapter, the aim is to investigate the role of the kinetics chosen for the ODE model of a biological process on the sensitivity properties of oscillating systems. First, in section 6.1, it is distinguished between reactions following Michaelis-Menten kinetics and mass action kinetics. Additionally, the notion of saturation is defined as used in this work for transient processes (section 6.1.3). Furthermore, it is explained how Michaelis-Menten kinetics is understood in this work and its effects on oscillatory properties described in literature are introduced (sections 6.1.4, 6.1.5). The second part deals with the effects of saturating kinetics on the sensitivity of steady states (section 6.2). Therein, the Goldbeter-Koshland switch for which zero-order ultrasensitivity has been described (Goldbeter and Koshland [1981]) is re-examined (section 6.2.3) and a similar analysis is performed for determining the sensitivity of the steady state of chain-like systems (sections 6.2.4, 6.2.5). Third, in section 6.3, the chain models are altered to employ Michaelis-Menten kinetics. Sensitivity analyses of these models (section 6.3.2) and a further investigation of their results (sections 6.3.4 to 6.3.5) are performed to reveal the effects of saturating kinetics on the period and amplitude sensitivities of oscillating systems. The final sections deal with the effect of saturation on the oscillation probability (section 6.4), and a summary and discussion of the obtained results (section 6.5).

6.1. Mass action kinetics vs. Michaelis-Menten kinetics and saturation

In biological models, the kinetics of a reaction indicates how the source species (or substrates) of a reaction enter into the reaction rate. Thereby, the temporal concentration change of the product species of the reaction and also of the source species if matter flow is present is decisively influenced.

6.1.1. Mass action kinetics

The most simple kinetics used in many biological models is a rate following the law of mass action:

$$\nu = k \cdot S_1 \cdot S_2. \tag{6.1}$$

Therein, the reaction rate is linearly proportional to each source species concentration (in this case S_1, S_2) with a rate coefficient k. The reaction rate gets larger, i.e. the reaction gets faster with increasing source species concentrations. Typically, this kinetics is chosen in models of signaling pathways as long as nothing is known about detailed mechanisms or special characteristics of a reaction (Klipp and Liebermeister [2006]).

6.1.2. Michaelis-Menten kinetics

Another highly employed kinetics has been suggested by Michaelis and Menten [1913]. The so-called Michaelis-Menten kinetics is originally derived from the simplification of an enzyme-driven reaction scheme as given in Figure 6.1. Before being converted to a product, the substrate is reversibly bound to an enzyme. Every single reaction is assumed to follow mass action kinetics in every source species. If assuming the total enzyme concentration to be constant and small compared to the substrate concentration, the quasi steady state assumption (QSSA) of the enzyme-substrate complex, $\frac{dES}{dt} = 0$, is valid and leads to a reaction rate of product formation of

$$\nu = k \cdot ES = V_{max} \frac{S}{S + K_M}. \tag{6.2}$$

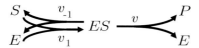

Figure 6.1.: Reaction scheme for the derivation of Michaelis-Menten kinetics. A substrate S binds reversibly to an enzyme E, the enzyme-substrate complex ES is converted to a product P.

The total enzyme concentration enters linearly into V_{max}, the rate coefficients of the three reactions enter into K_M. Details of the derivation of this equation are given in the Appendix, section A.4.1, and can be found in any textbook for enzyme kinetics (e.g. Fell [1997], Cornish-Bowden [2004]).

A reaction rate governed by Michaelis-Menten kinetics increases, i.e. the reaction gets faster, with increasing substrate concentration. Since the term $\frac{S}{S+K_M}$ is always smaller than one, the rate coefficient determines the maximal possible velocity of the reaction ν. That is why the rate coefficient of a reaction ν governed by Michaelis-Menten kinetics is most often denoted by V_{max}, also V_m or simply V as in this work (compare section 2.1.2).

6.1.3. Saturation

Reactions governed by Michaelis-Menten kinetics have one special property which distinguishes them from mass action reactions: their rates can saturate. In Figure 6.2, a Michaelis-Menten reaction rate $\nu = V\frac{S}{S+K_M}$ is given depending on the substrate concentration. For comparison, the reaction rate $\nu = k \cdot S$ of a mass action reaction with the same slope at $S = 0$ (being $\frac{\mathrm{d}}{\mathrm{d}t}V\frac{S}{S+K_M}|_{S=0} = V\frac{K_M}{(K_M+S)^2}|_{S=0} = \frac{V}{K_M}$) is also shown. While the mass action reaction rate depends always linearly on the substrate concentration and gets very large for high substrate concentrations, a Michaelis-Menten reaction rate exhibits two qualitatively different types of behavior for different substrate concentrations (Figure 6.2):

(i) The region of first-order kinetics (green). For low substrate concentrations, the Michaelis-Menten reaction rate depends linearly on the concentration. In this region, the reaction rate is nearly indistinguishable from a mass action reaction with rate coefficient $\frac{V}{K_M}$.

(ii) The region of zero-order kinetics (orange). For large substrate concentrations, the Michaelis-Menten reaction rate saturates approaching the rate coefficient (or maximal reaction velocity) V. Hence, the reaction rate gets independent from the substrate concentration and is almost constant.

How large or small the substrate concentration needs to be in order for the reaction to exhibit first-order kinetics or zero-order kinetics depends on the K_M-value of the reaction rate. In this work, a measure of the degree of saturation of a Michaelis-Menten reaction ν is the *saturation index* sat_ν^* for constant substrate concentration S. It is defined by

$$sat_\nu^* = \frac{S}{S + K_M} \tag{6.3}$$

according to a definition for steady states (Kurosawa and Iwasa [2002]). The index sat_ν^* takes values between zero (no saturation) and one (complete saturation).

With this index, the saturation of a reaction is easily determined for steady states and in other cases with constant substrate concentrations. Nevertheless, it is more complicated to define the saturation for transient states where the substrate concentration is varying as for example in oscillations. One method would be to define the saturation of a limit cycle oscillation via the saturation index of its steady state

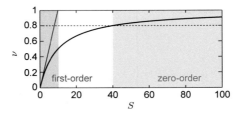

Figure 6.2.: Michaelis-Menten kinetics and saturation. A Michaelis-Menten kinetics reaction rate $\nu = V\frac{S}{S+K_M}$ is given in dependence on the substrate concentration with $V = 1$ and $K_M = 10$ (thick black curve). The thin solid line shows the mass action kinetics reaction rate $\nu = k \cdot S$ with a rate coefficient of $k = 1/10$ yielding the same slope for $S = 0$ as the Michaelis-Menten kinetics reaction. The region of first-order kinetics is shaded in green. Here, the reaction rate is nearly linear in the substrate. The region of zero-order kinetics is shaded in orange. Therein, the reaction rate is nearly independent from the substrate concentration. Concentration and time are given in arbitrary units.

(as done e.g. in Kurosawa and Iwasa [2002]). This practice has the disadvantage to not capture the true saturation characteristics of the reaction during the oscillation, especially if the steady state concentrations are higher or lower than the concentrations obtained during the oscillation. This is possible for ODE systems with more than two variables, for example shown in the bifurcation diagram of S_1 for the negative feedback chain model (section 4.4.2, Figure 4.5 A).

In Figure 6.3 A, two cycles of an oscillation of a substrate S are given together with the first- and zero-order regions for a Michaelis-Menten reaction depending on S. In panel B, the values of the saturation index as defined in Equation 6.3 of the Michaelis-Menten reaction for the substrate concentration at particular time points are shown. It can be seen that in this example, the Michaelis-Menten reaction rate switches from the first-order to the zero-order region of kinetics and back during the cycles of the oscillation. In the following, in order to keep it as simple as possible, for oscillatory processes only the largest saturation index obtained during the oscillation is considered. The *general saturation index* of a Michaelis-Menten reaction ν, sat_ν, is defined as the saturation index of the reaction rate at maximal substrate concentration:

$$sat_\nu = \frac{S_{max}}{S_{max} + K_M}. \tag{6.4}$$

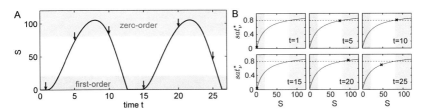

Figure 6.3.: Saturation index for oscillatory processes. In panel A, the concentration development of substrate S in time (both in arbitrary units) is shown for an oscillatory process. The areas of first- and zero-order kinetics are shaded for an example Michaelis-Menten reaction ν depending on S assuming $K_M = 20$ for ν. The panels in B indicate the saturation index of ν for time points $t = 1, 5, 10, 15, 20, 25$ during the cycle. The largest saturation index of ν obtained during the cycle is 0.84 for a substrate concentration of $S_{max} = 105$ which gives the general saturation index $sat_\nu = 0.84$.

121

With the help of this generalized measure for saturation, it is also possible to quantify the saturation property of reactions depending on a substrate which participates in transient processes.

Throughout this work, unless stated otherwise, a Michaelis-Menten reaction is termed to be *saturated* whenever its (general) saturation index is above 0.8. This coincides with the region where the slope of the reaction rate with respect to changes in substrate is less than 4% of the maximal obtained slope $1/K_M$ which arises at very low substrate concentrations. A (general) saturation index of one which signifies total independence of the reaction rate from the substrate is referred to as *complete saturation* or *total saturation*. However, this is only a theoretical value and cannot be reached practically in a Michaelis-Menten reaction in nature.

6.1.4. The meaning of Michaelis-Menten kinetics in this work

In general, Michaelis-Menten kinetics is assigned rather to reactions occurring in metabolic than in signaling networks (Klipp and Liebermeister [2006]) since Michaelis-Menten kinetics requires very small total enzyme concentrations, $E_{tot} \ll K_M + S_{tot}$. This may be the case for many metabolic enzymes (e.g. Albe et al. [1990]). In signaling pathways, however, the concentrations of enzymes and their substrates are often at similar orders of magnitude, for example in the MAPK cascade, and thus the use of Michaelis-Menten kinetics as simplification of a reaction scheme incorporating enzymes is debated (Blüthgen et al. [2006], Salazar and Höfer [2006]). There is an on-going discussion on whether and under which conditions the QSSA is valid, and which features of a dynamical system are lacking if a reaction is simplified according to it (e.g. Flach and Schnell [2006], discussed in section 6.5.2).

Throughout this work, Michaelis-Menten kinetics is employed as the most simple possible phenomenological description of a reaction which can get saturated in the regulating or source species. It is important to notice that Michaelis-Menten kinetics is not considered as simplification of an enzyme-driven reaction scheme here. The assumption made for the examination is that the QSSA underlying saturating reactions is valid, or at least that the simplification to a Michaelis-Menten kinetics is not affecting the examined behavior.

Saturating reactions occur whenever a separation of time-scales in a QSSA for reducing the complexity of a system is made (Flach and Schnell [2006]). Separation of time-scales may occur if binding to a possibly non-abundant other species, e.g. a metabolite, enzyme, micro RNA, or even ATP (Thron [1999]), is required for the reaction. It does not only occur in enzyme-driven reactions for which the Michaelis-Menten kinetics rate was initially designated. Separation of time-scales and simplification of multiple reactions steps to one reaction may be valid for production reactions as e.g. transcription where transcription factors (typically not being present in excess) have to bind and limit the rate, as well as for conversion or degradation reactions for which binding of the molecules to lowly abundant proteins e.g. phosphatases, kinases or components of a destruction complex (e.g. known from Wnt signaling, Kimelman and Xu [2006], Saito-Diaz et al. [2012]) are required. Saturation is even likely to occur in transport reactions where transporter molecules might be needed whose concentrations are limiting.

For simplicity, in this work, the species in which the reaction rate can saturate is still called substrate of the reaction although it may not be based on an enzyme-driven reaction scheme. The notions Michaelis-Menten kinetics and saturating kinetics are used synonymously in this work. Michaelis-Menten kinetics is hence utilized in the following as means to determine the effect of saturation on the sensitivity of the period and the amplitude of oscillations.

6.1.5. Known effects of saturating kinetics on oscillations

Saturating kinetics as Michaelis-Menten kinetics for degradation reactions are known to increase the probability of oscillations, i.e. they enlarge the parameter region in which sustained oscillations can be observed (Kurosawa and Iwasa [2002], Xu and Qu [2012]). In particular, saturated degradations have been shown to reduce the cooperativity required in the model to exhibit sustained oscillations (Goldbeter [2013]). For the period and amplitude of circadian oscillations, the degree of saturation of mRNA and protein degradation has been shown to have crucial effects on whether the period is increased or decreased in the response to changes in these degradation rates (Gerard et al. [2009]). Additionally, increased saturation in phosphorylation and dephosphorylation reactions has been shown to enlarge the amplitude of cell-cycle oscillations (Gerard et al. [2012]). Similar results have been obtained for the amplitude as well as period of minimal models of glycolytic and cell cycle oscillations which increases when employing non-linear instead of linear degradations (Goldbeter [2013]). Also the shape of the oscillation has been reported to depend on the kinetics of the degradation reactions (Goldbeter [2013]). The effect of saturating kinetics on the sensitivity of the period or amplitude of oscillations, however, remains largely unexplored.

In contrast, plenty is known about how saturation affects the sensitivity of the steady state (section 6.2). This constant entity of dynamical systems is much easier to examine than a transient behavior as in oscillations. Still, results from the investigation of saturation in steady state sensitivity may be valuable to explain certain observations concerning the sensitivity of period and amplitude in oscillations. Therefore, the starting point to analyze the effects of saturating kinetics will be an analysis of its impact on the steady state variability for particular, small systems (section 6.2). Afterwards, the role of saturating kinetics for oscillatory systems is examined with the help of the chain models (section 6.3).

6.2. The effect of saturating kinetics on steady state sensitivity

The effect of saturation on the sensitivity of steady state responses has subject to intensive research. However, despite the existence of multiple motifs for creating an ultrasensitive steady state (Zhang et al. [2013]), for example positive feedback and homo-multimerization, only one of the known basic motifs directly relies on saturating kinetics without requiring further cooperativity: the phenomenon of zero-order ultrasensitivity in conversion cycles. This principle has been found and established by Goldbeter and Koshland [1981].

In many publications, the subject of steady state sensitivities in conversion cycles has been investigated, including for example the influence of the spatial distribution of the participating enzymes (Van Albada and Ten Wolde [2007]), or the effect of proteins with multiple phosphorylation sites (Markevich et al. [2004], Salazar and Höfer [2009]). Also the fact that saturation in conversion cycles can results in bistability has been reported frequently (e.g. Markevich et al. [2004], Thron [1999]).

Besides, the steady state of cascades of conversions cycles, which are considered to form the fundamental structure of MAPK signaling (Blüthgen and Legewie [2008], Kholodenko and Birtwistle [2009], Plotnikov et al. [2011]), has been subject to sensitivity studies (e.g. Goldbeter and Koshland [1984], Markevich et al. [2004], Blüthgen et al. [2006], Blüthgen [2006], Li and Srividhya [2010], Fritsche-Guenther et al. [2011]). Also conditions under which zero-order ultrasensitivity in conversion cycles is reduced have been described (discussed in section 6.5.2).

The ultrasensitivity experienced in the context of saturating kinetics is based on the examinations by Goldbeter and Koshland [1981]. In the following, the ideas and results of this publication are revised and extended to other system which are not conversion cycles, but resemble chain-like systems. First, zero-order ultrasensitivity of the steady state is introduced for the Goldbeter-Koshland switch (section 6.2.1). Their measure for steady state sensitivity, the response coefficient, is described (section 6.2.2) and used to reconstruct the complete characteristic for the Goldbeter-Koshland switch (section 6.2.3). The same measure is applied to study the steady state sensitivity of two systems derived by step-wise alteration of the switch: First, adding a production and a degradation reaction to the switch creating an open switch (section 6.2.4), and second, leaving out the backward reaction of the switch creating a short chain (section 6.2.5). Like this, the effect of saturation on the steady state sensitivity of a system resembling a part of a linear chain model is explored. Results obtained in this analysis may be also applicable to the sensitivity of oscillatory properties in the chain model.

6.2.1. Zero-order ultrasensitivity in the Goldbeter-Koshland switch

In their publication, Goldbeter and Koshland [1981] have mathematically analyzed the steady state behavior of a system composed of a species S_1 which is reversibly converted by enzyme-driven reactions to another species S_2 (reaction scheme in Figure 6.4 A). This system can be interpreted as, for example, the description of a protein switching from unmodified to modified state and back e.g. by a phosphorylation and a dephosphorylation, respectively. If assuming to capture each process in one saturating reaction, i.e. in a Michaelis-Menten kinetics reaction, the following

ODE system is determining the concentrations of S_1 and S_2:

$$\frac{dS_1}{dt} = V_2 \frac{S_2}{S_2 + K_2} - V_1 \frac{S_1}{S_1 + K_1}$$
$$\frac{dS_2}{dt} = V_1 \frac{S_1}{S_1 + K_1} - V_2 \frac{S_2}{S_2 + K_2}. \tag{6.5}$$

For reaction ν_i, the parameters are the maximal reaction velocity V_i and the K_M-value K_i, $i = 1, 2$. This system is also known as the Goldbeter-Koshland switch. In the system, there is no production or degradation included, hence the total protein concentration $S_{tot} = S_1 + S_2$ is constant. For fixed S_{tot} and for each combination of positive parameters, the unique positive steady state can be calculated analytically (expression given in the Appendix, Equation A.3).

Goldbeter and Koshland have theoretically examined how the steady state of the system behaves for changes in maximal reaction velocities V_i. If changing the quotient of the maximal reaction velocities V_1/V_2 from low to high, the steady state of S_1 changes from high to low (Figure 6.4 B). The authors have observed that for intermediate K_M-values (green curve in Figure 6.4 B), this steady state transition is moderately steep. On the contrary, whenever both reactions have a low K_M-value compared to the total protein concentration (red curve in Figure 6.4 B), the steady

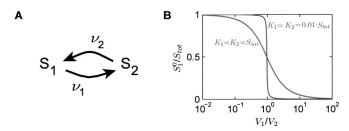

Figure 6.4.: The Goldbeter-Koshland switch. A: Model scheme. Species S_1 and S_2 are interchanged via enzyme-driven reactions ν_1 and ν_2 whose kinetics are simplified to Michaelis-Menten rates, see the ODEs in Equations 6.5. B: Development of the steady state percentage S_1^0/S_{tot} in dependence on the maximal reaction velocity ratio V_1/V_2 and the K_M-values. Figure contents are similar to Figure 1 in Goldbeter and Koshland [1981]. In the green curve, the general saturation index of both reactions is $sat_{\nu_i} = 0.5$, for the red curve, it is $sat_{\nu_i} = 0.99$.

state changes abruptly with small variations of the quotient V_1/V_2. Goldbeter and Koshland have termed this behavior *zero-order ultrasensitivity*. If the reactions are in a zero-order regime with respect to the substrate, i.e. they are saturated reactions, the steady state is more sensitive than the "Michaelis-Menten response to stimulus", i.e. the change in the steady state is steeper than the change of a Michaelis-Menten reaction rate when changing its substrate concentration.

6.2.2. Sensitivity of the steady state and ultrasensitivity: The response coefficient R_ν

In order to define the sensitivity of the steady state which is expressed in the steepness of the steady state transition, Goldbeter and Koshland have defined the response coefficient

$$R_\nu = \frac{(V_1/V_2)_{0.1}}{(V_1/V_2)_{0.9}} \tag{6.6}$$

as a ratio of particular maximal reaction velocity ratios. Thereby, $(V_1/V_2)_{0.1}$ (resp. $(V_1/V_2)_{0.9}$) is the reaction velocity ratio necessary to obtain a steady state with 10% (resp. 90%) of the total protein being S_1, i.e. $S_1^0 = 0.1 S_{tot}$ (resp. $S_1^0 = 0.9 S_{tot}$). For example, having a response coefficient of five means that for the nine-fold change of the steady state concentration (from $0.1 S_{tot}$ to $0.9 S_{tot}$) a five-fold change of the maximal reaction velocity ratio is necessary.

The lower the response coefficient is, the more sensitive the steady state behaves in response to maximal reaction velocity alterations. According to the definition in Goldbeter and Koshland [1981], the steady state is ultrasensitive as soon as $R_\nu < 81$, since an R_ν of 81 is obtained in the curve where a Michaelis-Menten reaction is given in dependence on the substrate concentration (Figure 6.2 in section 6.1.3), because the substrate concentration has to change by 81-fold in order to increase the reaction rate from $0.1 \cdot V_{max}$ to $0.9 \cdot V_{max}$. According to this definition, also a steady state depending linearly on the reaction rate would be ultrasensitive since trivially, $R_\nu = 9 < 81$ is obtained therein. Consequently, in this work, ultrasensitivity of the steady state is instead considered only for $R_\nu < 9$, i.e. for the steady state being more sensitive than a linear stimulus-response curve (compare discussion in Legewie et al. [2005]). For a response coefficient of $R_\nu = 1$, the steady state concentration is switching instantaneously without any variation of the maximal reaction velocities. This behavior, which is only theoretically possible, is referred to as *infinite ultrasensitivity* (Goldbeter and Koshland [1981]).

Indeed, the response coefficient can be defined for any ODE system. One needs to choose the species S of which the steady state concentration S^0 shall be examined and define a threshold concentration value S_{thres} for this species. Then, one has to choose the reaction rate V (or reaction rate ratio as for example V_1/V_2) towards which the steady state sensitivity shall be calculated. V_a, $a = 0.1, 0.9$, is defined as the reaction rate (or reaction rate ratio) for which it is $S^0 = a \cdot S_{thres}$. For consistency, define the response coefficient to be larger than one by

$$R_\nu = \begin{cases} V_{0.1}/V_{0.9} & \text{if } V_{0.1} > V_{0.9} \\ V_{0.9}/V_{0.1} & \text{if } V_{0.1} \leq V_{0.9}. \end{cases} \tag{6.7}$$

In this generalized definition, the response coefficient and thus the steady state sensitivity may depend on the choice of the threshold concentration. However, this is not the case for the systems examined here (sections 6.2.4, 6.2.5).

In fact, the response coefficient R_ν characterizes the global behavior of the concentration for changes in the rate coefficients. Similarly, as a local measure also the response coefficient

$$R_V^{S^0} = \frac{\partial \log S^0}{\partial \log V} = \frac{\partial S^0}{\partial V} \frac{V}{S^0} \tag{6.8}$$

from MCA (Kacser and Burns [1973], Heinrich and Rapoport [1974]) could be used for the estimation of the steady state sensitivity. It is the analogue for infinitesimal perturbations to the sensitivity coefficients employed in this work (compare Equations 2.5 in section 2.2). The response coefficient $R_V^{S^0}$ yields the relative change of the steady state S^0 to (infinitesimal) relative changes in V and is used frequently for characterizing steady state behavior (e.g. in Szedlacsek et al. [1992], Ortega et al. [2002], Legewie et al. [2005], Blüthgen et al. [2006], Zhang et al. [2013]). In contrast to R_ν, $R_V^{S^0}$ can also take negative values (if increasing V decreases S^0), and the steady state is more sensitive the larger $|R_V^{S^0}|$ is. In accordance with the definition of ultrasensitivity via R_ν, the steady state of S is considered to be ultrasensitive if $|R_V^{S^0}| > 1$ (compare also Legewie et al. [2005], Zhang et al. [2013]), since $|R_V^{S^0}| = 1$ for a linear dependency of S^0 on V. Infinite ultrasensitivity is obtained for $|R_V^{S^0}| = \infty$.

If investigating the maximal $R_V^{S^0}$ for the steady state varying in an interval $S^0 \in [S_{min}, S_{max}]$, this maximal response coefficient also delivers a global steady state sensitivity characteristics and makes comparison with the global measure R_ν more appropriate. In fact, concerning the sensitivity of the steady state and its dependency on saturation properties, it is not surprising that similar results are obtained for $R_V^{S^0}$ and R_ν in the examined models (compare the Appendix, section A.4.3, to the results in the following sections 6.2.3, 6.2.4, 6.2.5).

6.2.3. Re-examination of the Goldbeter-Koshland switch

Goldbeter and Koshland [1981] have examined only a limited number of saturation property combinations of the two conversion reactions and included only the argumentation about the steady state transition steepness for both rates getting saturated. It is not shown what happens if only one of the reactions acts in the zero-order regime. Therefore, the Goldbeter-Koshland switch is re-examined.

Thereby, if the steady state concentration S_1^0 is known, the other steady state concentration is easily obtained by $S_2^0 = S_{tot} - S_1^0$. Hence, the maximal reaction velocity ratio necessary to obtain $S_1 = 0.1 \cdot S_{tot}$ (resp. $S_1 = 0.9 \cdot S_{tot}$) is the same as that for $S_2 = 0.9 \cdot S_{tot}$ (resp. $S_2 = 0.1 \cdot S_{tot}$). Consequently, also the response coefficients and steady state sensitivities for S_1 and S_2 are identical. Additionally, since the Goldbeter-Koshland switch is symmetrical, regarding S_1^0 for changes in V_1/V_2 is exactly identical to examining S_2^0 for changes in V_2/V_1. Consequently, in the following, without loss of generality, the analysis is restraint to the development and steady state sensitivity of S_1 with regard to changes in V_1/V_2.

The general saturation index definition from section 6.1.3 is used for reactions ν_1 and ν_2. In every steady state curve over varying V_1/V_2, S_1^0 approaches S_{tot} for $V_1/V_2 \to \infty$ and zero for $V_1/V_2 \to 0$ (see the Appendix, section A.4.2, Equation A.3). The same holds true for $S_2^0 = S_{tot} - S_1^0$ with reversed limits for V_1/V_2. Therefore, the general saturation indexes of the reactions are given by $sat_{\nu_i} = \frac{S_{tot}}{S_{tot} + K_i}$. In Figure 6.5, a selection of steady state developments of S_1 for a fixed saturation index of reaction ν_1 (panel A) or fixed saturation index of reaction ν_2 (panel B) are given. Note that the scale for V_1/V_2 is logarithmic to facilitate comparison of fold-changes

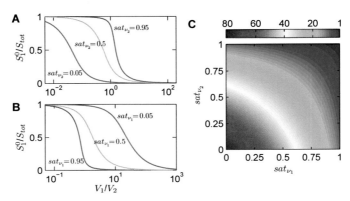

Figure 6.5.: Re-examination of the steady state sensitivity of the Goldbeter-Koshland switch. A, B: Steady state development of S_1 for fixed general saturation index of reaction ν_1, $sat_{\nu_1} = 0.75$, and varying general saturation indexes of sat_{ν_2} (A) or fixed $sat_{\nu_2} = 0.75$ and varying sat_{ν_1} (B). C: Response coefficients R_ν for the steady state of S_1 for the Goldbeter-Koshland switch in dependence on the general saturation indexes of reactions ν_1 and ν_2.

of V_1/V_2 for different orders of magnitude which matter for the response coefficient. It is obvious that as soon as one of the reactions gets more saturated, the steepness of the steady state transition, i.e. the sensitivity of the steady state, increases. This behavior is also reflected in the values of the response coefficients R_ν measuring the steepness of the steady state transition (depicted in color-code in Figure 6.5 C). The response coefficient depending on the saturation indexes of the reactions is calculated via

$$R_\nu = \frac{81(1 - 0.9sat_{\nu_1}) \cdot (1 - 0.9sat_{\nu_2})}{(1 - 0.1sat_{\nu_2}) \cdot (1 - 0.1sat_{\nu_1})} \tag{6.9}$$

The derivation of this equation is given in the Appendix, section A.4.2. Note that the response coefficient does not depend explicitly on the total amount of protein present in the system, S_{tot}. It only enters implicitly via the general saturation indexes of the reactions.

Figure 6.5 C shows that whenever one of the reactions gets saturated, the response coefficient gets smaller, i.e. the sensitivity of the steady state towards changes in maximal reaction velocity ratio increases. If one of the reactions approaches complete saturation with a general saturation index of approximately one and the other acts entirely in the first-order region ($sat_{\nu_i} \approx 0$), the response coefficient decreases to a value close to nine, which corresponds to the sensitivity of a linear response. Lower values, i.e. higher sensitivities, are obtained whenever both reactions approach complete saturation. For both reactions being close to saturated, i.e. $sat_{\nu_1} = sat_{\nu_2} = 0.8$, a response coefficient of 7.5 is obtained. This means that a change of 7.5 fold of the maximal reaction velocity ratio is necessary to shift the steady state of S_1 of the system from 90% of S_{tot} to 10% of S_{tot}. Infinite ultrasensitivity, which is obtained for a response coefficient of one, would be theoretically reached if both reactions were completely saturated ($sat_{\nu_1} = sat_{\nu_2} = 1$). All in all, the steady state sensitivity increases as soon as one reaction acts in the zero-order regime in the Goldbeter-Koshland switch. If both reactions get more saturated, the effect is more pronounced and the steady state even becomes ultrasensitive.

6.2.4. Open switch: The Goldbeter-Koshland switch with production and degradation

Different scenarios have been suggested which could even increase the sensitivities, as cascades of conversion cycles (e.g. Goldbeter and Koshland [1984]), or which could decrease the observed sensitivities in conversion cycles (discussed in section 6.5.2). However, the question which has not been addressed or clarified yet is if the enhanced sensitivity of the steady state in the zero-order regime observed in the Goldbeter-Koshland switch also occurs if the system is no longer a conversion cycle, i.e. if the sum of the participating species is not constant. Therefore, in the following, an altered Goldbeter-Koshland switch is investigated for steady state sensitivity: an "open switch" where a production reaction ν_0 and a degradation reaction ν_3 are included in the Goldbeter-Koshland switch (Figure 6.6 A). For a constant substrate production, the validity of the quasi steady state assumption for a single reaction has been investigated and proven to be correct for sufficiently high production rates (Stoleriu et al. [2004]). In the following analysis, it is assumed that for both reactions ν_1 and ν_2, the use of saturating Michaelis-Menten kinetics is valid. For simplicity, the production is constant, $\nu_0 = k_0$, and the degradation reaction is governed by mass action kinetics, $\nu_3 = k_3 \cdot S_2$. The ODE system for the open switch is the following:

$$\frac{dS_1}{dt} = k_0 - V_1 \frac{S_1}{S_1 + K_1} + V_2 \frac{S_2}{S_2 + K_2}$$

$$\frac{dS_2}{dt} = V_1 \frac{S_1}{S_1 + K_1} - V_2 \frac{S_2}{S_2 + K_2} - k_3 S_2.$$

(6.10)

In contrast to the original Goldbeter-Koshland switch, a short calculation reveals that in the open switch the steady state concentration of S_2 does not depend on the reaction rates of the conversion cycles, but only on the production and degradation. It is $S_2^0 = \frac{k_0}{k_3}$. The saturation index for reaction ν_2 is hence given by $sat_{\nu_2} = \frac{S_2^0}{S_2^0 + K_2}$. Since S_2^0 remains constant for any changes in V_1 or V_2, the examination of the steady state transition is limited to the observation of S_1^0. The steady state concentration of S_1 can be calculated analytically (derived in the Appendix, section A.4.2). In dependence on V_1, K_1 and $k_0 + V_2 sat_{\nu_2}$, it is given by

$$S_1^0 = K_1 \frac{k_0 + V_2 sat_{\nu_2}}{V_1 - (k_0 + V_2 sat_{\nu_2})}.$$

Note that in contrast to the Goldbeter-Koshland switch, the open switch is not symmetrical in V_1 and V_2 and thus relative perturbations of V_1 have different effects than perturbations of V_2 on S_1^0, and the response coefficients R_{ν_1} or R_{ν_2} for perturbations only in V_1 or only in V_2, respectively, need to be considered separately.

For decreasing V_1, the steady state concentration of S_1 increases deliberately, and for increasing V_1, it decreases to zero (see Figure 6.6 B). Hence, in order to define a measure of saturation to compare the curves, a threshold value S_{thres} is fixed. The maximal saturation index obtained during a steady state change up to the threshold value is $sat_{\nu_1} = \frac{S_{thres}}{S_{thres} + K_1}$.

Now, the steady state transition steepness for changes in maximal reaction velocities, i.e. the steady state sensitivity, of the open switch is examined. In Figure 6.6 B, the development of S_1^0 up to a threshold of $S_{thres} = 1000$ for varying V_1 and different saturation properties of the reactions is shown. For fixed general saturation index $sat_{\nu_1} = 0.75$ (Figure 6.6 B left), the steady state transitions appear to be similar for all variations of the general saturation indexes of ν_2 indicating similar steady state sensitivities. In contrast, for fixed $sat_{\nu_2} = 0.75$ (Figure 6.6 B right), an increase in the general saturation index of reaction ν_1 leads to increased transition steepness and hence increased sensitivity of the steady state. The response coefficients R_{ν_1} of the steady state of S_1 given a threshold value of S_{thres} for changes in V_1 can help verify that the steady state sensitivity towards changes in V_1 depends only on the saturation of ν_1. It is given by

$$R_{\nu_1} = \frac{(V_1)_{0.1}}{(V_1)_{0.9}} = 9\frac{1 - 0.9sat_{\nu_1}}{1 - 0.1sat_{\nu_1}} \qquad (6.11)$$

which is derived by straightforward calculation (performed in the Appendix, section A.4.2). From Equation 6.11, it follows that indeed the response coefficient and hence the steady state sensitivity is independent from the saturation index of reaction ν_2. Additionally, it does not depend on the explicit value of the threshold S_{thres}, nor on any other parameter of the system except for the general saturation index of ν_1. The response coefficient R_{ν_1} in dependence on sat_{ν_1} and sat_{ν_2} is given in Figure 6.6 C. As already derived from the equation, the steady state sensitivity does not depend on sat_{ν_2}, but only on the saturation of ν_1. R_{ν_1} decreases with increasing saturation of ν_1 and takes values between nine and one. The steady state in this situation is thus always ultrasensitive with regard to changes in V_1.

The response coefficient yielding the steady state sensitivity of S_1 with respect to changes in V_2, R_{ν_2}, is based on the change from a minimal concentration plus ten percent of the possible response up to the threshold concentration, $S_{min} + 0.1 \cdot (S_{thres} - S_{min})$ to $S_{min} + 0.9 \cdot (S_{thres} - S_{min})$. The reason for this is that for varying V_2 and fixed values of the other parameters, the steady state concentration of S_1 is only reaching down to $S_{min} > 0$. The development of S_1^0 in dependence on V_2 for varying saturation levels of ν_1 or ν_2 but otherwise fixed parameters is shown in Figure 6.6 D. Note that sat_{ν_1} alters S_{min} and thus the reference lines, between which the change in S_1^0 is considered, are altered from curve to curve in the left panel. It seems as if, again, only the change in saturation level of ν_1, but not of ν_2, influences the steady state sensitivity. Indeed, the response coefficient R_{ν_2} is given by

$$R_{\nu_2} = \frac{(V_1)_{0.9}}{(V_1)_{0.1}} = 9\frac{\left(\frac{sat_{\nu_1}}{1 - sat_{\nu_1}} - \frac{k_0}{V_1 - k_0}\right) \cdot 0.1 + 1}{\left(\frac{sat_{\nu_1}}{1 - sat_{\nu_1}} - \frac{k_0}{V_1 - k_0}\right) \cdot 0.9 + 1}. \qquad (6.12)$$

Note that R_{ν_2} is only defined for $\frac{k_0}{V_1 - k_0} < sat_{\nu_1}/(1 - sat_{\nu_1})$, because otherwise $S_{min} \geq$

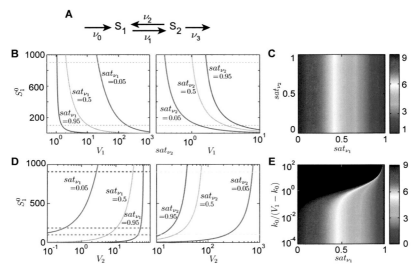

Figure 6.6.: Steady state sensitivity of the open switch. A: Model scheme for the open switch which is the Goldbeter-Koshland switch with constant production and linear degradation. The conversion reactions ν_1 and ν_2 are governed by Michaelis-Menten kinetics. For the set of differential equations, see Equations 6.10. B: Development of the steady state of S_1 for varying V_1 up to a threshold concentration of 1000 for fixed $sat_{\nu_2} = 0.75$ (left) or fixed $sat_{\nu_1} = 0.75$ (right) and varying saturation of the respective other reaction rate. For these curves, it is $k_0 = 0.25$ and $V_2 = 1$. Dotted horizontal lines indicate $0.1 \cdot S_{thres}$ and $0.9 \cdot S_{thres}$ levels. C: Response coefficients R_{ν_1} for different general saturation indexes of the conversion reactions ν_1 and ν_2 given in Equation 6.11. D: Development of the steady state of S_1 for varying V_2 up to a threshold concentration of 1000 for fixed $sat_{\nu_2} = 0.75$ (left) or fixed $sat_{\nu_1} = 0.75$ (right) and varying saturation of the respective other reaction rate. For these curves, it is $k_0 = 0.25$ and $V_1 = 50$. Dotted curves in the according colors (left) or gray (right) indicate $0.1 \cdot (S_{thres} - S_{min}) + S_{min}$ and $0.9 \cdot (S_{thres} - S_{min}) + S_{min}$ levels. E: Response coefficients R_{ν_2} for different general saturation indexes of the conversion reaction ν_1 and different values of $k_0/(V_1 - k_0)$ using Equation 6.12. The regions with $k_0/(V_1 - k_0) \geq sat_{\nu_1}/(1 - sat_{\nu_1})$, where the response coefficient is not defined for all threshold concentrations, are depicted in black. Note that R_{ν_2} is independent from sat_{ν_2}.

S_{thres}. The derivation of this equation and the condition is shown in the Appendix, section A.4.2. According to Equation 6.12, interestingly, the steady state sensitivity of S_1 does not depend on sat_{ν_2} or the considered threshold concentration. Instead, it is influenced by sat_{ν_1} and $k_0/(V_1 - k_0)$. The values of R_{ν_2} are given in Figure 6.6 E. For increasing saturation of ν_1, the response coefficient gets smaller and thus the steady state is more sensitive towards changes in V_2. The response coefficient gets larger, i.e. the steady state gets less sensitive the larger $k_0/(V_1 - k_0)$ is. R_{ν_2} takes values between nine and one, thus S_1^0 is nearly always ultrasensitive with respect to changes in V_2.

The comparison of the steady state sensitivities of the open switch to those of the original Goldbeter-Koshland switch (section 6.2.3) indicates that the production and degradation reactions eliminate the influence of the saturation status of one of the reactions of the conversion cycle (ν_2) on the steady state sensitivity of S_1. Indeed, the equation for the response coefficient of R_{ν_1} is even the same as that for the original Goldbeter-Koshland switch for ν_2 being completely saturated, i.e. $sat_{\nu_2} = 1$, irrespective of the actual saturation index of ν_2. Therefore, the response coefficient in the open switch reaches values as low as those obtained in the original Goldbeter-Koshland switch, but not as high as observed therein. As soon as ν_1 is saturated, i.e. $sat_{\nu_1} > 0.8$, the response coefficient R_{ν_1} of the open switch takes values smaller than 2.8. This means that for one reaction being saturated (ν_1), it is sufficient to change the maximal reaction velocity ratio by less than 2.8-fold in order to decrease the steady state by nine-fold from 90% to 10% of an arbitrary threshold concentration. Recalling that for the Goldbeter-Koshland switch without production and degradation, a fold-change of 7.5-fold is necessary to achieve the same reduction if *both* reactions satisfy $sat_{\nu_i} = 0.8$, one can consider the open switch to have an even larger steady state sensitivity imposed mainly by the saturation properties of one reaction, ν_1.

6.2.5. Short chain: The open switch without backward reaction

In the preceding part, ultrasensitivity in the zero-order regime has been proven for the open switch model. In other biological processes, though, there might not occur a switch-like structure, but reactions may be practically irreversible and lack their opposing reaction. Here, the open switch is altered further and examined with respect to steady state sensitivity: An open switch lacking the backward-reaction ν_2. This leads to a chain-like model referred to as "short chain" here which is composed of two species (Figure 6.7 A). The ODE system for this model is given by

$$\begin{aligned}
\frac{\mathrm{d}S_1}{\mathrm{d}t} &= k_0 - V_1 \frac{S_1}{S_1 + K_1} \\
\frac{\mathrm{d}S_2}{\mathrm{d}t} &= V_1 \frac{S_1}{S_1 + K_1} - k_2 S_2.
\end{aligned} \tag{6.13}$$

The steady state of the system can be calculated analytically. Since $S_2^0 = \frac{k_0}{k_2}$, the steady state concentration of S_2 is, as in the open switch, not affected by the maximal reaction velocity V_1 or the saturation of reaction ν_1. Therefore, the analysis is restricted to the sensitivity of the steady state transition of S_1 towards changes in V_1. The steady state of S_1 depends only on the production rate k_0, the maximal reaction velocity V_1 and the K_M-value K_1, and is given by

$$S_1^0 = \frac{k_0 K_1}{V_1 - k_0}$$

(calculation provided in the Appendix, section A.4.2). In Figure 6.7 B, the development of S_1^0 in dependence on changes in the maximal reaction velocity V_1 for fixed $k_0 = 1$ is shown. The steady state concentration can increase deliberately with decreasing V_1. Therefore, again a threshold value S_{thres} is fixed and the general saturation index of reaction ν_1 is defined by $sat_{\nu_1} = \frac{S_{thres}}{S_{thres} + K_1}$.

With regard to steady state transition steepness, Figure 6.7 B is revisited: Comparing the steady state curves for low and high saturation of reaction ν_1 (green versus red curve), the latter shows a decisively steeper increase than the first. To verify this observation of enhanced sensitivity for zero-order kinetics, the response coefficient R_ν are calculated for the steady state of S_1, a threshold value of S_{tres} and for changes in V_1. Thereby, $(V_1)_a$ determines the reaction velocity V_1 necessary to yield a steady state concentration for S_1 of $a \cdot S_{thres}$. The response coefficient is given by

$$R_\nu = \frac{(V_1)_{0.1}}{(V_1)_{0.9}} = 9 \frac{1 - 0.9 sat_{\nu_1}}{1 - 0.1 sat_{\nu_1}} \tag{6.14}$$

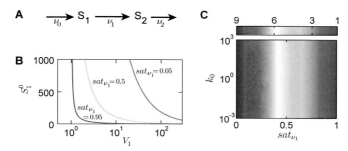

Figure 6.7.: Steady state sensitivity of the short chain. A: Model scheme for the short chain which is the open switch omitting the backward reaction. Only the conversion reaction ν_1 is governed by Michaelis-Menten kinetics. For the set of differential equations, see Equations 6.13. B: Development of the steady state of S_1 for varying V_1 up to a threshold concentration of 1000 for fixed $k_0 = 1$ and varying saturation index sat_{ν_1}. C: Response coefficients for different saturation indexes of the conversion reaction ν_1 and different production rates k_0 as given in Equation 6.14.

(equation derived in the Appendix, section A.4.2). The equation is exactly the same as that for the open switch (Equation 6.11). The reason for this could lie in the fact that already for the open switch, the saturation of ν_2 does not influence on the sensitivity of the steady state of S_1 towards changes in V_1. Hence, a system lacking ν_2 could be expected to display a similar steady state sensitivity as the open switch.

For the sake of completeness, the response coefficients in dependence on the saturation index of ν_1 and the production rate k_0 are depicted in Figure 6.7 C. The response coefficients are independent from the production rate (as well as from the explicit threshold value S_{thres}, seen from the formula). Again, the response coefficient yields a maximal value of nine if the general saturation index of reaction ν_1 is zero. For an increasing general saturation index of ν_1 the response coefficient decreases. This means the more ν_1 is governed by a zero-order regime, the more sensitive the steady state gets with respect to changes in V_1. The maximal degree of the steady state sensitivity is again similar to that obtained for the Goldbeter-Koshland switch: the response coefficient approaches values close to one (infinite sensitivity) for ν_1 approaching complete saturation.

This is the most striking result: The system examined here can, like the original

Goldbeter-Koshland switch, reach infinite (i.e. very high) sensitivity despite the fact that only *one* reaction with saturating kinetics occurs. Thus, one can conclude that for enhanced sensitivity, the system needs not to be a conversion cycle, but one possibly saturating reaction is sufficient. The more this reaction is saturated, the stronger the steady state of its source species changes for varying maximal reaction velocity. The steady state of the system reveals to be more sensitive for the Michaelis-Menten reaction acting in the zero-order regime than in the first-order regime.

6.2.6. Enhanced sensitivity of the steady state

The preceding examination shows that enhanced sensitivity for the zero-order regime is not restricted to switch-like structures with constant substrate sums, but also occurs for systems with production and degradation (e.g. the open switch) or systems only exhibiting a single reaction governed by saturating kinetics (e.g. the short chain). Thereby, however, saturated conditions do not directly imply that one obtains a steep steady state change, but the observed steady state change also depends on the original values and amount of change of the maximal reaction velocities. This can be observed for example in the classical Goldbeter-Koshland switch in Figure 6.5 A and B: Even though both reactions are either saturated or close to saturation in the red curves, a small change in V_1/V_2 results only in minor changes in the steady state of S_1 if V_1/V_2 is very small or very large. Consequently, kinetics in a zero-order regime do not guarantee that one observes increased steady state sensitivity for every change in maximal reaction velocities, but they rather give rise to an increased probability to observe high sensitivities.

6.3. The effect of saturating kinetics on the period and amplitude sensitivities of the chain model

In the preceding part (section 6.2), it has been shown that in simple systems resembling parts of the chain model, saturating reaction kinetics can enhance the sensitivity of the steady state. In this section, it is thoroughly examined how saturating kinetics influence the sensitivity of the period and amplitude of oscillating systems with the help of the chain model of length four. Therefore, first, the chain model is modified to employ Michaelis-Menten kinetics and sensitivity analyses are performed (section 6.3.2). Secondly, an examination of the sensitivity coefficients of the parameters reveals the role of the K_M-values which are only introduced by changing the kinetics and do not occur in the chain model with mass action kinetics (section 6.3.3). Third, the results of the sensitivity analyses are used to investigate how the number of saturated reactions in general influences the period and amplitude sensitivity (section 6.3.4). Fourth, the influence of the position of saturated reactions on period and amplitude sensitivity is analyzed (section 6.3.5). Finally, the effect of saturation on the oscillation probability is discussed (section 6.4).

6.3.1. Michaelis-Menten kinetics in the chain model

In order to compare the effect of feedback type and matter flow on the sensitivity of the chain models in chapters 4 and 5, the simplest scenario has been assumed and therefore mass action kinetics has been employed. Now, the influence of employing saturating kinetics is investigated by examining the chain models with negative or positive feedback introduced in section 4.2.

The seven non-constant reactions of the chain models are altered towards employing saturating, i.e. Michaelis-Menten, kinetics. The production reaction ν_1 of S_1 remains constant. In the seven other reactions $\nu_2, \ldots, \nu_8$, a Michaelis-Menten kinetics term replaces the linear species term. The K_M-values are referred to as $K_2, \ldots, K_8$, the index accords with that of the reaction the K_M-value occurs in. In the following, also the notation of the rate coefficients is changed. In agreement with the notation used in biological literature for Michaelis-Menten terms, they are referred to as V_i, $i = 2, \ldots, 8$ and called maximal reaction velocities since they exhibit this function in saturating kinetics (sections 2.1.2, 6.1.2). The reaction where the feedback enters, ν_2, is a product of the maximal reaction velocity V_2, a Michaelis-Menten term depending on S_1 and the feedback term as defined in section 4.2.1. The remaining six reactions are governed by Michaelis-Menten kinetics depending on the source species of the reaction. The number of nl-parameters of the models compared to the models with mass action kinetics increases from one (the inhibition constant of the Hill-term only) to eight (having seven K_M-values of the Michaelis-Menten reactions). The ODE systems for the chain models with saturating kinetics are given in the Appendix, section A.7.4. Matter flow is assumed to occur for all reactions from one species to another, i.e. reactions ν_2, ν_4 and ν_6 are conversion reactions. The effect of altered matter flow properties is discussed in section 6.5.1.

Employing saturating kinetics in the chain models results in an alteration of their stability properties towards increased occurrence of unstable steady states (Appendix, section A.4.4). The rate laws of the reactions need to be sufficiently nonlinear to destabilize the steady state (Novak and Tyson [2008]). Destabilization of the steady state is apparently fostered by the additional source of non-linearities when employing saturating kinetics. In fact, a Hill coefficient of $n = 1$ is sufficient for the chain model with either feedback type, which complies with very low cooperativity and very low feedback strength (compare section 4.2.2) to induce unstable steady states and also oscillations. The results of the sensitivity analyses examining period and amplitude sensitivities for the chain models with Hill coefficient $n = 1$ are compared to those obtained for the original Hill coefficients ($n = 9$ and $n = 2$ for negative and positive feedback, respectively) in the Appendix (section A.4.5).

6.3.2. Sensitivity analysis of the chain models with linear or saturating kinetics

For the chain models with saturating kinetics and negative or positive feedback, sensitivity analyses are performed. The results are shown in Figure 6.8 and in the Appendix (Tables A.14, A.15). All in all, introduction of saturating kinetics instead of linear mass action kinetics into the chain model leads to a crucial change in both the period and the amplitude sensitivity characteristics.

In Figure 6.8 A, the chain models with negative feedback and different kinetics are compared. For the model with Michaelis-Menten kinetics, higher period and amplitude sensitivities can be observed as well as a decisively broadened period sensitivity distribution.

First, the details for the period sensitivity distribution changes are described. The median period sensitivity is decisively and significantly increased by 3.5-fold from 0.19 to 0.66 for the model with saturating kinetics (Table A.14, MWU p-value is

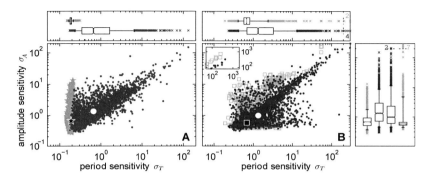

Figure 6.8.: Impact of the kinetics on the sensitivity of the chain model. A: Amplitude and period sensitivities for the negative feedback chain model with mass action kinetics (orange stars, black star: median values) or saturating kinetics (dark red dots, white circle: median values). B: Amplitude and period sensitivities for the positive feedback chain model with mass action kinetics (light green squares, black square: median values) or saturating kinetics (dark green dots, white circle: median values). The inset displays parameter sets with sensitivity values > 80. The box-plots in the according colors capture the distribution characteristics of the period sensitivities (top) and amplitude sensitivities (right).

zero, Table A.15). The 95th percentile of the period sensitivity distribution of the chain model with negative feedback and mass action kinetics is even smaller than the fifth percentile of the period sensitivity distribution of the chain model with negative feedback and saturating kinetics (values of 0.22 and 0.23, respectively, Table A.14). This means that 95% of the parameter sets for the model with saturating kinetics have a larger period sensitivity than 95% of the model with mass action kinetics. Also the 90%R and hence the dispersion of the period sensitivity distribution is strongly increased by 126-fold from 0.05 to 6.32 if introducing Michaelis-Menten kinetics.

Similar observations can be made for the amplitude sensitivity distributions of the negative feedback chain model. If employing saturating kinetics instead of mass action kinetics, the median amplitude sensitivity rises significantly by two-fold from 0.66 to 1.34 (Table A.14, MWU p-value $1.6 \cdot 10^{-207}$, Table A.15). Also the 90%R as measure of dispersion of the amplitude sensitivity distributions is increased if applying Michaelis-Menten kinetics. Its value changes by nearly six-fold from 1.49 for the model using mass action kinetics to 8.54 for the model with saturating kinetics (Table A.14).

Furthermore, if applying Michaelis-Menten kinetics instead of mass action kinetics, the correlation between amplitude sensitivity and period sensitivity is strengthened. Spearman's rank correlation coefficient R_S rises from 0.30 for mass action kinetics to 0.76 for saturating kinetics (Table A.14).

For the positive feedback chain models, similar observations as for the negative feedback chain models can be made for employing saturating kinetics (Figure 6.8 B). The period sensitivity distribution is significantly shifted to higher values for the model with Michaelis-Menten kinetics. The median period sensitivity enlarges by a factor of two from 0.68 to 1.33 (Table A.14, MWU p-value $3.8 \cdot 10^{-173}$, Table A.15). If using saturating kinetics instead of mass action kinetics, the dispersion of the period sensitivity distribution of the chain model with positive feedback rises by a factor of nearly 6.5 yielding 90%R values of 1.99 for mass action kinetics and 12.92 for saturating kinetics (Table A.14).

The amplitude sensitivity distribution is significantly increased for the positive feedback chain model if employing saturating kinetics. The median value rises by 1.7-fold from 0.57 for mass action kinetics to 0.98 for saturating kinetics (Table A.14, MWU p-value $1.2 \cdot 10^{-175}$, Table A.15). The dispersion of the amplitude sensitivity distribution as indicated by the 90%R enlarges by a factor of 2.5 from 4.48 to 11.28 for the model with saturating kinetics (Table A.14).

As for the chain model with negative feedback, introducing saturating kinetics increases the correlation between amplitude sensitivity and period sensitivity in the chain model with positive feedback. Low correlation with $R_S = 0.42$ for the model with mass action kinetics is replaced by moderate correlation with $R_S = 0.64$ for the model with saturating kinetics (Table A.14).

Altogether, using saturating kinetics enlarges both the period and amplitude sensitivities of the chain model and broadens the sensitivity distributions of the chain model with either type of feedback. Further examinations in the following sections try to unravel the origin for this increase in sensitivity and try to draw parallels to the examination of the steady state sensitivities from section 6.2.

6.3.3. Influence of the K_M-values

One possible reason for the increase in the sensitivities of the chain models with saturating kinetics compared to the chain models with mass action kinetics could be the introduction of the seven K_M-values as seven additional nl-parameters. The impact of the K_M-values on period and amplitude sensitivity is examined by investigating the sensitivity coefficients of the parameters of the model with saturating kinetics and comparing them to those of the model with mass action kinetics for either type of feedback (Figure 6.9).

For each feedback type, the K_M-values reveal to be systematically less influential on the period and amplitude than their respective maximal reaction velocities since the latter consistently exhibit bigger median values of their $|R^T|$ as well as $|R^A|$ (comparing second and third box-plot for reactions ν_2-ν_8 in Figure 6.9 A, B for negative feedback, C, D for positive feedback). This result can be easily explained. If applying a percentage change in the maximal reaction velocity V of a reaction rate $\nu = V \frac{S}{S+K_M}$ with saturating kinetics, the reaction rate ν is changing by exactly the same percentage, irrespective of the actual parameter values or species concentration. In contrast, if changing the K_M-value of reaction ν by a certain percentage, the actual change in the reaction rate is not the same but depends on the saturation of the reaction. How the reaction rate changes upon perturbation of K_M depending on the saturation index of the reaction is shown in Figure 6.10. Indeed, the more the reaction is saturated, the less the reaction rate is altered for applied changes in the K_M-value. Therefore, since the reaction rate is changed at most similarly, but more frequently less by the same perturbation in the K_M-value than in the maximal reaction velocity, also the period and amplitude change less for the K_M-values. Consequently, the sensitivity coefficients of the K_M-values are smaller than those of the according maximal reaction velocities.

From this, one can immediately conclude that the increase in period sensitivities and amplitude sensitivities observed for employing saturating kinetics instead of linear kinetics cannot be attributed solely to the introduction of additional parameters, the K_M-values. Instead, all maximal reaction velocities display larger median $|R^T|$ and median $|R^A|$ than the according rate coefficients from the model with mass action kinetics (see Figure 6.9 A-D, except for slightly smaller values for ν_2 in panel

B and ν_3 in panel C), also for the constant reaction ν_1 whose kinetics has not been altered. Thus, employing saturating instead of linear kinetics increases the sensitivities of period and amplitude with respect to perturbations of the rate coefficients in oscillating systems. This is in agreement with the results from the analysis of the steady state sensitivity in section 6.2. Therein, it has been shown that the sensitiv-

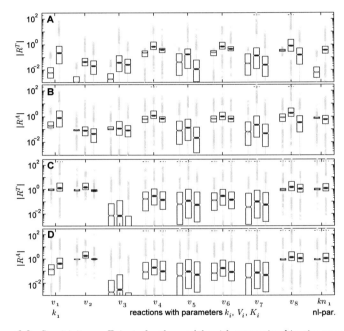

Figure 6.9.: Sensitivity coefficients for the models with saturating kinetics compared to the models with mass action kinetics for negative feedback (panels A, B) or positive feedback (panels C, D). The absolute period sensitivity coefficient $|R^T|$ distributions (panels A, C) or absolute amplitude sensitivity $|R^A|$ distributions (panels B, D) are shown as box-plots. In each panel, for each reaction, the results for the rate coefficient of the reaction for the model with mass action kinetics (median given in orange, light green), for the maximal reaction velocity and, if applicable, for the K_M-value of the reaction for the model with saturating kinetics (median given in dark red, dark green) are given next to each other to allow for direct comparison.

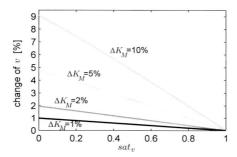

Figure 6.10.: Relative change of a Michaelis-Menten reaction rate $\nu = V \frac{S}{S+K_M}$ for relative changes ΔK_M in the K_M-value of $1\%, 2\%, 5\%, 10\%$. Constant maximal reaction velocity V and constant substrate concentration S are assumed. Results are depicted depending on the saturation index sat_ν of the reaction. Note that for increasing K_M, the Michaelis-Menten term decreases.

ity of the steady state with respect to changes in the maximal reaction velocities is increased if reactions are saturated. Obviously, this is also the case for the sensitivity of properties of oscillating systems if reactions can potentially saturate.

6.3.4. The influence of the number of saturated reactions

In section 6.3.2 it has been shown that saturating kinetics can increase the period and amplitude sensitivities of the chain models. For systems with a structure comparable to parts of the chain model, the steady state sensitivity has been shown to increase with increasing level of saturation, i.e. saturation index, of Michaelis-Menten reactions (section 6.2). Accordingly, in this section, it is elaborated how the sensitivity of period and amplitude is affected by the level of saturation of a reaction.

For this purpose, the 2500 parameter sets generated during the sensitivity analysis of the chain model with Michaelis-Menten kinetics and negative or positive feedback (section 6.3.2) are re-examined with respect to saturation properties (Figure 6.11). The chain models with saturating kinetics possess seven reactions being governed by Michaelis-Menten kinetics, ν_2-ν_8, which can hence be saturated or not. Recall that a Michaelis-Menten reaction rate is termed saturated if its general saturation index is above 0.8 (section 6.1.3). For each parameter set, the number of saturated reactions is counted. The numbers of parameter sets obtained in each category – from zero up to seven reactions being saturated – are depicted in Figure 6.11 A for the negative feedback chain model and D for the positive feedback chain model. For each category the period sensitivity distributions (Figure 6.11 B, E) and amplitude sensitivity distributions (Figure 6.11 C, F) are given by box-plots. The median values are emphasized by circles.

For the period sensitivities (Figure 6.11 B and E), a clear trend of increasing sensitivity values for increasing number of saturated reaction rates can be identified for both chain models. This is reflected in a perfect positive correlation between the number of saturated reactions and the median period sensitivities (values of $R_S = 1$

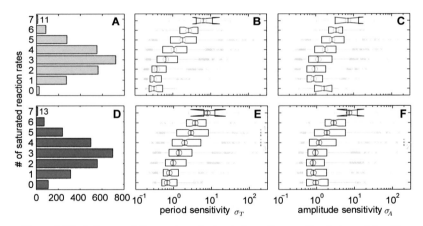

Figure 6.11.: Influence of the number of saturated reactions on the chain models with saturating kinetics. A reaction ν is considered to be saturated whenever its general saturation index satisfies $sat_\nu > 0.8$. A and D: Number of parameter sets obtained for each number of saturated reactions for the negative and positive feedback chain model, respectively. Numbers < 20 are given next to the bar. Note that numbers sum up to 2500 for each panel. B and C: Period and amplitude sensitivity distributions, respectively, for each number of saturated reactions for the negative feedback chain model. Median values are emphasized with circles. E and F: Period and amplitude sensitivity distributions, respectively, for each number of saturated reactions for the positive feedback chain model.

Table 6.1.: Correlations between saturation level and sensitivities measured by Spearman's rank correlation R_S for the chain models with saturating kinetics and negative feedback (fb.) or positive feedback. Given are the correlations between the number of saturated reactions (# sat rct) and the according median sensitivity (μ) or the sensitivities of all parameter sets (σ), as well as the according p-values (compare section 2.4.5). Data corresponds to Figure 6.11.

correlation between		negative fb.				positive fb.			
		period	p-val	ampl.	p-val	period	p-val	ampl.	p-val
# sat rct vs.	μ	1	0.0072	0.86	0.0176	1	0.0072	0.83	0.0210
# sat rct vs.	σ	0.54	$9.7 \cdot 10^{-161}$	0.36	$1.0 \cdot 10^{-62}$	0.24	$1.8 \cdot 10^{-33}$	0.39	$1.0 \cdot 10^{-84}$

for both models, Table 6.1).

For the amplitude sensitivities (Figure 6.11 C, F), a slight decrease of median sensitivities from zero up to one (negative feedback) or two (positive feedback) saturated reaction rates can be observed. The median sensitivities then increase with increasing number of saturated reactions. Still, the correlations between the number of saturated reactions and the median amplitude sensitivity are strong (values of 0.86 and 0.83, Table 6.1).

The overall observed tendencies in Figure 6.11 suggest that it is not only the potential to saturate which enhances the sensitivities for the chain models yielding Michaelis-Menten kinetics in contrast to the model with linear kinetics as observed in Figure 6.8. Instead, also the saturation level, i.e. the number of reactions actually being saturated, matters for period and amplitude sensitivities. Note that the correlation between saturation levels and sensitivity is still positive but decisively smaller when comparing numbers of saturated reactions with the sensitivity of all parameter sets (second line in Table 6.1). This indicates that high saturation levels of reactions do not automatically result in increased sensitivities of oscillatory properties but rather in an increased probability to yield high sensitivities (as for steady states, section 6.2.6). The results obtained in Figure 6.11 are reproduced for different definitions of a saturated reaction, i.e. for $sat_\nu > a$ with $a \neq 0.8$ (Appendix, section A.4.6).

6.3.5. Influence of particular reactions being saturated

In the the preceding section 6.3.4, the number of saturated reactions is found to be crucial for the period and amplitude sensitivities but it is not characterized which of the reactions are saturated. Thus, it is now examined whether the saturation of particular reactions of the chain models have stronger impact on the sensitivity of the oscillatory properties than others. As a first step, for each reaction, the sensitivities of the parameter sets without the reaction being saturated are compared to the sensitivities obtained for the parameter sets with the reaction being saturated (Figure 6.12, Table 6.2).

Concerning the period sensitivities, having any reaction saturated significantly increases the period sensitivities compared to this reaction not being saturated for both chain models (Figure 6.12, medians and MWU p-values in Table 6.2). Reactions ν_3, ν_4 or ν_6 getting saturated for the negative feedback chain model increases the period sensitivity most (fold-changes of the medians of 1.9, 3.8, and 3.6, respectively, calculated from Table 6.2). For the positive feedback chain models, the saturation level of reaction ν_2 is affecting the median period sensitivity most (fold-

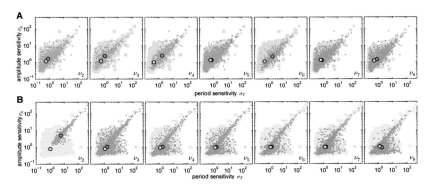

Figure 6.12.: Influence of the saturation of particular reactions on the sensitivities of the chain models with saturating kinetics. Each panel gives the sensitivities for the negative (A) and positive (B) feedback chain model, respectively, for the parameter sets where reaction ν_i indicated in the lower right is non-saturated (light orange or light green, circle for median values) compared to those with ν_i being saturated (dark orange or dark green, circle for median values).

Table 6.2.: Results for the medians and MWU test comparing the sensitivity distributions of the chain models for a reaction (rct) being non-saturated (nsat) to the reaction being saturated (sat). Given are the medians (μ), the MWU p-values (MWU p), U, z-values (z-val) and the number of parameter sets obtained in each category (par. sets). Data corresponds to Figure 6.12.

rct	period sensitivity					amplitude sensitivity					par. sets	
	μ nsat	μ sat	MWU p	U	z-val	μ nsat	μ sat	MWU p	U	z-val	nsat	sat
negative feedback chain model with saturating kinetics												
ν_2	0.57	0.79	$3.1 \cdot 10^{-13}$	-7.2	643309	1.16	1.63	$5.1 \cdot 10^{-11}$	-6.5	656421	1383	1117
ν_3	0.59	1.10	$8.3 \cdot 10^{-20}$	-9.0	346264	1.17	2.38	$1.2 \cdot 10^{-33}$	-12.0	304181	2035	465
ν_4	0.44	1.67	$2.3 \cdot 10^{-142}$	-25.4	247761	1.00	2.46	$3.5 \cdot 10^{-82}$	-19.2	351723	1716	784
ν_5	0.59	0.69	0.0056	-2.5	247761	1.33	1.34	0.44	-0.15	608896	667	1833
ν_6	0.50	1.78	$6.0 \cdot 10^{-98}$	-21.0	211682	1.13	2.25	$2.7 \cdot 10^{-36}$	-12.5	336085	1972	528
ν_7	0.61	0.71	$7.0 \cdot 10^{-6}$	-4.3	668243	1.36	1.32	0.3558	-0.37	738287	980	1520
ν_8	0.50	0.79	$6.6 \cdot 10^{-26}$	-10.5	568665	1.21	1.43	$7.8 \cdot 10^{-5}$	-3.8	687081	1017	1483
positive feedback chain model with saturating kinetics												
ν_2	1.01	5.52	$1.5 \cdot 10^{-179}$	-28.5	79887	0.79	5.05	$8.4 \cdot 10^{-206}$	-30.6	50799	2018	482
ν_3	1.03	1.47	$1.1 \cdot 10^{-11}$	-6.7	528449	0.81	1.06	$5.0 \cdot 10^{-11}$	-6.5	532173	714	1786
ν_4	1.13	1.67	$6.6 \cdot 10^{-14}$	-7.4	625133	0.95	1.05	0.0156	-2.2	718363	1472	1028
ν_5	1.22	1.48	$1.6 \cdot 10^{-7}$	-5.1	687536	0.95	1.02	0.0022	-2.8	728428	1195	1305
ν_6	1.26	1.49	$1.7 \cdot 10^{-6}$	-4.6	587703	0.97	1.01	0.0626	-1.5	639563	1732	768
ν_7	1.25	1.41	$7.1 \cdot 10^{-5}$	-3.8	686002	0.97	1.00	0.0661	-1.5	726720	1486	1014
ν_8	1.18	1.74	$1.9 \cdot 10^{-13}$	-7.3	577825	1.04	0.90	$2.2 \cdot 10^{-5}$	-4.1	632130	1648	852

change of median value 5.5), the saturation levels in reactions ν_3, ν_4, and ν_8 yield the next largest influence (fold changes of median values 1.4, 1.5, and 1.5, respectively, from non-saturated to saturated).

The amplitude sensitivities are significantly increased for all reactions switching from non-saturated to saturated reaction rates except for ν_5 and ν_7 for the negative feedback chain model. Reactions ν_4, ν_6 and ν_7 do not have a significant influence for the positive feedback chain model (compare MWU p-values and medians in Table 6.2). Saturation in ν_8 even decreases the median amplitude sensitivity compared to ν_8 not being saturated (Table 6.2). The strongest increases in median values are observed again for changing saturation in ν_3, ν_4, and ν_6 for the negative feedback chain model (two-, 2.5-, and two-fold increase in medians, Table 6.2) and ν_2 and ν_3 for the positive feedback chain model (6.4- and 1.3-fold increase in medians, Table 6.2).

Hence, how much the period sensitivities as well as amplitude sensitivities increase, if at all, for saturated reactions compared to the reactions being non-saturated depends on the particular reaction and model. A more detailed analysis of the influence of the saturation of the less influential reactions on the sensitivities for constant saturation level of the more influential reactions is shown in the Appendix, section A.4.7. Therein, for the negative feedback chain models, if keeping the saturation of reactions ν_3, ν_4, ν_6 the same, altering the saturation of reaction ν_7 does not reveal significant changes in period sensitivities or amplitude sensitivities, while changing the saturation of ν_2 only significantly alters the period sensitivities under certain conditions. Additionally, under three of the examined eight conditions (for only ν_6 being saturated, ν_3 and ν_4 being saturated or ν_4 and ν_6 being saturated), the saturation levels of the other reactions ν_2, ν_5, ν_7 and ν_8 do not significantly influence the sensitivities. For the positive feedback chain model, if ν_2 is not saturated, the saturation of all other reactions significantly increases the period sensitivities. In the case of ν_2 being saturated, only altering reaction ν_4 from non-saturated to saturated significantly alters (increases) the period sensitivity. The amplitude sensitivities are only significantly affected by altering the saturation of ν_3 and ν_8, and only if ν_2 is not saturated.

Additionally, if examining the chain models in which not all seven reactions but only a subset employs saturating kinetics (Appendix, section A.4.8), effects of combinations of reactions with saturating kinetics can be observed. This is in particular the case for the negative feedback chain model. Therein, if only all degradation reactions or only all conversion reactions exhibit saturating kinetics, the period and amplitude sensitivities are, if at all, only slightly increased. In contrast, if all seven reactions are allowed to saturate in the negative feedback chain model by using saturating kinetics, a very strong increase in period as well as amplitude sensitivities is observed.

To sum it up, general conclusions on the effects of the saturation of single reactions or combinations of reactions are difficult to draw and each case needs to be examined separately in order to derive precise conclusions regarding changes in period and amplitude sensitivities due to saturating kinetics.

6.4. Saturation and oscillation probability

As elaborated in section 6.3.5, saturating kinetics can have different effects depending on the particular reaction and on the combinations of reactions whose kinetics are considered. Some of those combinations of saturated reactions even may be favored and occur more often in the course of sampling because they lead more often to oscillations, some may occur only rarely because they reduce the oscillation probability.

Indeed, there is a known correlation between saturation properties and oscillation probability, at least for circadian oscillation models.Saturation properties of particular reaction steps in a three- and four-variable circadian oscillation model and their effect on the stability of the steady state and hence the oscillation probability have been thoroughly examined (Kurosawa and Iwasa [2002]). It has been found that saturation in any of the reactions being part of the feedback loop suppresses oscillations, saturation of degradation reactions and back-transport of protein to cytosol (a reaction reverse to the loop direction) makes oscillations more likely to occur. In contrast to the saturation index defined in this thesis, in the published analysis, the saturation index of a Michaelis-Menten kinetics reaction has been defined as being the quotient $\frac{S^0}{S^0 + K_M}$, where S^0 is the steady state concentration of the participating species. Kurosawa and Iwasa [2002] have summed up that the effect of saturation on the oscillation probability depends on the location of the saturated reaction within the network, and that the saturation index is small for inner-loop-reactions, i.e. reactions forming the feedback loop, and high for branch reactions, i.e. reactions not forming the feedback loop as degradation rates or reactions in reverse direction to the feedback loop. That the saturation of degradation rates enhances the occurrence of oscillations has also been stated for different models (Xu and Qu [2012], Goldbeter [2013]).

The results of the analysis of circadian models by Kurosawa and Iwasa [2002] are confirmed for the 2500 sets of the chain model with negative feedback and saturating kinetics. In Figure 6.13 A, for each reaction ν_i, $i = 2, \ldots, 8$, the number of parameter sets for which the reaction is saturated (black), or non-saturated (white) is determined. Both numbers sum up to 2500 for each reaction. The degradation rates of the inner-loop species (ν_5, ν_7 and ν_8) are more often saturated than non-saturated.

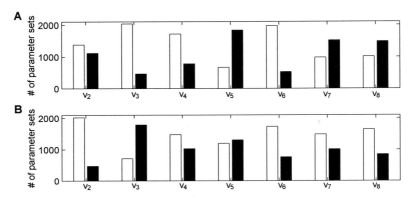

Figure 6.13.: Influence of position of saturation on oscillation probability. Examined is the chain model with saturating kinetics and negative (A) or positive (B) feedback. The number of sets are counted for which the Michaelis-Menten term of the indicated reaction ν_i is saturated (black) or not (white). For each reaction, numbers sum up to 2500.

Contrariwise, the degradation of the outer-loop species S_1 and the inner-loop conversions (ν_3, ν_4 and ν_6) are more often non-saturated than saturated. Indeed, the possibility of saturation for all reactions via introduction of Michaelis-Menten kinetics enlarges the oscillation probability for the whole model by one order of magnitude from 0.013% for the model employing mass action kinetics only to 0.18% (Table 6.3).

In contrast, for the positive feedback chain model the oscillation probability decreases by more than one order of magnitude from 1.67% for the mass action kinetics model to 0.15% when introducing Michaelis-Menten kinetics (Table 6.3). This trend is also reflected when examining the relationship between the saturation of few reactions and oscillation probability for the positive feedback chain model and Michaelis-Menten kinetics (Figure 6.13 B). Except for the degradations of the first two species in the chain (ν_3 and ν_5), all reactions are more often non-saturated than saturated. This is particularly pronounced for ν_2, for which saturation occurs in only a fifth of the parameter sets yielding oscillations. This finding may be explained by the action of the negative effect being necessary for the occurrence of oscillations: Reaction ν_2 mediates the negative effect of S_4 on the concentration of

Table 6.3.: Oscillation probability of the chain models with negative feedback (neg. fb) or positive feedback (pos. fb) with Michaelis-Menten (MM) kinetics for a different subset of reactions. The oscillation probability is calculated as 2500 divided by the number of parameter sets sampled for the model.

MM in	neg. fb	pos. fb
no reaction	0.013%	1.67%
all reactions	0.18%	0.15%
degradations $(\nu_3, \nu_5, \nu_7, \nu_8)$	0.68%	1.73%
inner-loop conversions (ν_4, ν_6)	0.0026%	0.78%
all conversions (ν_2, ν_4, ν_6)	0.0032%	0.12%

S_1 towards the species S_2, S_3, S_4. In the case of ν_2 being saturated, this reaction turns widely independent from S_1. Consequently, the negative effect of S_1 on the rest of the chain and hence the negative feedback is weakened resulting in a smaller probability to oscillate (compare effect of feedback strength on steady state stability in section 4.2.2, Figure 4.2.2). This effect is not observed for the negative feedback chain model (Figure 6.13 A) since in this model, the negative feedback loop only comprises S_2 - S_4, and whether reaction ν_2 depends strongly or not on S_1 does not make any difference for the negative feedback.

For determining the effects of degradations, inner-loop conversions and the outer-loop conversions separately in more detail, the chain models with only a part of the seven reactions yielding saturating kinetics are analyzed. The obtained oscillation probabilities during the sensitivity analysis are given in Table 6.3, the resulting sensitivities are discussed in the Appendix, section A.4.8. Except for saturation in ν_2, the effects of the different reaction classes yielding saturating kinetics on the oscillation probability coincide for the negative feedback chain model and the positive feedback chain model. In particular, in accordance with the results from the literature (Kurosawa and Iwasa [2002], Xu and Qu [2012], Goldbeter [2013]), saturating instead of linear kinetics in the degradation reactions (ν_3, ν_5, ν_7, ν_9) increase the oscillation probability (only slightly for the positive feedback chain model, by 50-fold for the negative feedback chain model, Table 6.3). The finding that the effects for the positive feedback chain model are only small might be due to antagonistic effects of different degradations on the oscillation probability as suggested from Figure 6.13 B. In contrast to the degradations, saturating kinetics compared to linear kinetics in the inner-loop conversions (ν_4, ν_6) decisively hamper the occurrence of sustained oscillations (to less than half and one fifth for the positive and negative feedback chain model, respectively, Table 6.3). Saturating kinetics in ν_2, ν_4 and ν_6 lead to increased oscillation probabilities compared to saturating kinetics only in ν_4 and ν_6 in the negative feedback chain model (increase by one fifth, Table 6.3), but decisively decreased oscillation probability in the positive feedback chain model (decreased to one sixth, Table 6.3).

Altogether, the oscillation probability is decisively affected by the kinetics employed. In particular, for the negative feedback chain model, the oscillation probability can be drastically increased by introducing saturating kinetics. However, it depends on the particular reactions in the context of the examined model whether employing saturating kinetics renders the occurrence of sustained oscillations less or more probable.

6.5. Summary and discussion: The effect of saturating kinetics on sensitivity

In this chapter, the effects of saturating kinetics on steady state sensitivities of particular small models (section 6.2) as well as period and amplitude sensitivities of the chain models (section 6.3) are examined. The analysis relies on the use of Michaelis-Menten kinetics as phenomenological implementation of possibly saturating reaction kinetics (section 6.1). Concerning the steady state sensitivities, for an open switch and a short chain, it is sufficient to yield saturation in only one particular conversion reaction to obtain ultrasensitivity (section 6.2). Saturation also affects the sensitivities of oscillatory properties in the chain models (section 6.3). In fact, both, the positive feedback chain model as well as the negative feedback chain model, exhibit increased period and amplitude sensitivities when employing saturating kinetics in all reactions except for the constant production ν_1 (i.e. reactions ν_2 - ν_8, section 6.3.2). Thereby, in accordance with the results from the steady state sensitivities, indeed both period and amplitude sensitivities are increased towards changes in the maximal reaction velocities. Perturbations in the K_M-values lead to weaker alterations in period or amplitude compared to similar perturbations in the according maximal reaction velocities because the K_M-values alter their according reaction rates to a smaller extent (section 6.3.3). The period and amplitude sensitivities of both chain models increase with increasing number of saturated reactions (section 6.3.4), but the specific reactions which mainly influence the period and amplitude sensitivities for getting saturated differ between the two models (section 6.3.5). The kinetics and saturation level of some reactions play only a minor role for the period and amplitude sensitivities, but combinations of reactions yielding saturating kinetics can yield a more than additive effect on the sensitivities. Additionally, the oscillation probability is also affected by the saturation of the reactions. In particular, saturating kinetics in degradations increase the occurrence of sustained oscillations for both chain models (section 6.4).

Those findings are an extension of earlier analyses by Tsai et al. [2008]. Therein, only the feedback type has been held responsible for the sensitivity of period and amplitude. A fixed period and tunable amplitude has been found for all types of negative feedback oscillators, and a tunable period and partially fixed amplitude when adding a positive feedback. Here, the analyses show that also the kinetics used play a major role. Employing Michaelis-Menten kinetics decisively disturbs the property of fixed period for negative feedback oscillators and leads to oscillations with a variable period and amplitude even when lacking a positive feedback. This surprising issue is further pursued for a model proposed in Tsai et al. [2008] in section 7.1.

6.5.1. Saturating kinetics in conversions vs. saturating kinetics in regulated productions

In the examination of the steady state sensitivities (section 6.2), it is found that saturation in conversion reactions can be decisive for the sensitivities of the steady states. However, whether all and, if not all, which particular conversion reactions play a role depends on the examined system. The Goldbeter-Koshland switch requires both conversions to be able to saturate in order to exhibit ultrasensitivity of the steady states (section 6.2.3) whereas the saturation of one of the conversions can drive the steady state ultrasensitive in the open switch while the saturation of the other conversion does not play a role (section 6.2.4).

Similar observation can be made for the chain models with saturating kinetics in all seven reactions (section 6.3.5). For both models, the saturation of conversion reactions is most decisive in terms of period and amplitude sensitivities, but which exactly depends on the feedback type. Changes in saturation of the conversions ν_4 and ν_6 alter the sensitivities of the negative feedback chain model most whereas saturation of the conversion ν_2 is most decisive for the sensitivities of the positive feedback chain model.

Matter flow properties could alter these observations, as they can eliminate the decreasing effects of the conversions. In the chain models, the effect of matter flow properties and saturation can be assessed. Therefore, the sensitivities of the chain models with saturating kinetics, but lacking matter flow between species are examined (Appendix, section A.4.9). First of all, also in these models, the use of saturating kinetics increases the period and amplitude sensitivities, although less than for the models yielding more conversions. Analyses of the influence of single reactions show that the regulated productions ν_4 and ν_6 in the negative feedback chain model exhibit still the major influence for changes in their saturation level. Not surprisingly since it remained a conversion, also the saturation of reaction ν_2 is still the most influential for the positive feedback chain model. Thus, in contrast to assumptions derived from the steady state analysis where the saturation of the reaction acting only as production on the examined species yields no influence (section 6.2.4), the saturation levels of regulated productions are not decisively less influential than saturation levels of the according conversions on the period and amplitude sensitivities *per se*. However, lacking matter flow between species is found to enhance or reduce the susceptibility of the period and amplitude towards changes in the saturation of particular reactions (Appendix, section A.4.9).

6.5.2. Possible extensions of the examination of kinetics

In this chapter, Michaelis-Menten kinetics are employed as phenomenological description of a saturating reaction. Originally, it has been derived from an enzyme-kinetic scheme (Figure 6.1). Its validity is often discussed, but remains a major simplification in many enzyme-kinetic textbooks.

However, for conditions under which the validity of the QSSA as prerequisite for the simplification to Michaels-Menten kinetics is not fulfilled, reductions in the ultrasensitivity of the steady state have been observed. For example, if employing the total QSSA (Borghans et al. [1996]) as more broadly valid simplification instead of Michaelis-Menten kinetics, a loss of zero-order ultrasensitivity of the steady state in conversion cycles has been described (e.g. Ciliberto et al. [2007], Pedersen and Bersani [2010]). Zero-order ultrasensitivity of the steady state is also lost for conversion cycles of bifunctional enzymes, i.e. enzymes which catalyze both conversion reactions (Ortega et al. [2002], Shinar et al. [2009]). Contrariwise, it has been theoretically shown that if two distinct binding sites are present on the bifunctional enzyme, zero-order ultrasensitivity in conversion cycles can be retained (Straube [2013]). Product dependency of one of the conversions (Ortega et al. [2002]), non-productive binding (Goldbeter and Koshland [1981]) and sequestration effects for comparable enzyme and substrate concentrations (Blüthgen et al. [2006]) can also reduce the experienced steady state ultrasensitivity in conversion cycles.

These findings imply that for complete enzyme-driven schemes instead of the Michaelis-Menten kinetics simplification, the observed steady state sensitivities could be reduced. The particular effects of these types of network motifs on the sensitivity of oscillatory properties may be also interesting to examine.

In addition, there are more complex possibilities to describe a reaction rate which may saturate, as for example Hill-like terms in which exponents to the species and K_M-value occur. Indeed, the steady state stability in signaling cascade has been shown to behave differently for Michaelis-Menten kinetics compared to Hill kinetics (Qu and Vondriska [2009]), and thus also the sensitivities of period and amplitude may be affected. Furthermore, other mechanisms to alter the degradation reaction kinetics could be of interest for the sensitivities of period and amplitude. For example, a degradation following two different kinetic regimes depending on the time has been proposed and examined (Wong et al. [2007]). In many bacteria, a stronger degradation of monomers than homodimers has been observed (termed "cooperative stability", Buchler et al. [2005]) which results in nonlinear degradation kinetics different to the Michaelis-Menten or Hill regimes. Finally, in case of multiple substrates, other possibilities of saturation and of binding schemes can occur, for example in dependence on the binding order of the enzyme-substrate complexes (e.g. Fell [1997], Cornish-Bowden [2004]). The influence of these different kinetic regimes on the sensitivities of oscillatory properties remain to be elucidated as prospective work.

7. Further applications

In the previous chapters, the role of the feedback type, matter flow between species and kinetics has been examined with the help of the chain models. In chapter 7, models distinct from the chain models are investigated and the results obtained in the preceding chapters are applied. First, the basic oscillator models published by Tsai and colleagues [2008] are revisited. In their publication, the authors have investigated the influence of feedback loops on the tunability of oscillatory properties using a method different to the approach proposed in this thesis. A sensitivity analysis as introduced in this thesis is performed for some of the models proposed in Tsai et al. [2008] and the influence of kinetics in these models is examined (section 7.1). Second, with the help of exemplary models published for circadian oscillations and calcium oscillations, it is investigated to which extent the model structure can be inferred from the sensitivity results for these models using the insights from the previous chapters (section 7.2). Third, the results of the first two sections 7.1 and 7.2 are summarized and discussed, and further possible applications are shortly described as outlook: suggesting particular model characteristics as feedback types or kinetics to obtain a specific sensitivity, or using the sensitivity analysis to determine potential functions of oscillations (section 7.3).

7.1. It is not only feedback that matters

In this section, it is assessed how the work by Tsai and colleagues [2008] on the sensitivity of period and amplitude fits into the results obtained in this thesis. In particular, the role of kinetics in some of their proposed models is analyzed. The work of Tsai and colleagues [2008] is cited frequently ($\approx$220 citations according to Thomson Reuters Web of Science at the beginning of the year 2014) in different contexts: if discussing about the cell cycle (as the work has started the discussion with a cell cycle model), but mainly for the effect of the feedback type on the sensitivity of the period and amplitude. The approach used to estimate sensitivity in their work is different from the method introduced in this thesis. In their approach, only one particularly chosen parameter has been varied over its whole range of oscillatory behavior, and the therein occurring changes of period and amplitude (of one particular species) have been traced. As an example, for the period, the "operational frequency range", which is the range of periods obtained when changing the one parameter, is determined. The larger the range, the larger the so-called "tunability" of the period. The tunability coincides in part with the sensitivity used in this thesis, the more tunable the period or amplitude, the larger the sensitivity can be.

Additionally, different basic oscillator models than the chain models have been used by Tsai et al. [2008]. The structure of their original model with negative feedback is given in Figure 7.1 A. In fact, the model is derived from three conversion cycles in which a species S_i is converted to another species S_i^* and back, their sum is thereby constant, $S_i + S_i^* = 1$. S_1 positively regulates the conversion from S_2 to S_2^*, S_2 activates the conversion from S_3 to S_3^*, and S_3 the conversion from S_1 to S_1^*. If replacing the dependent variables $S_i^* = 1 - S_i$, the examined ODE system yields the regulatory structure in Figure 7.1 A (ODE system given in the Appendix, section A.7.6). A second model proposed in Tsai et al. [2008], herein referred to as the model with positive feedback, is derived from the model with negative feedback by adding a further conversion reaction from S_1^* to S_1 being positively regulated by S_1 (Figure 7.1 B, ODE system in the Appendix section A.7.6). This establishes a positive autoregulatory feedback, and the model with positive feedback has been used in the publication to compare the effects of feedback types on tunability or sensitivity.

Despite the different approach in the present work, in which (i) every single parameter is perturbed, (ii) the average of the obtained alteration in the period and in the mean of all amplitudes of the species is considered and (iii) the chain models are analyzed, the results concerning negative versus positive feedback in the work of Tsai et al. [2008] and this thesis are similar: The positive feedback model yields a less sensitive amplitude but a more sensitive period than the model with negative

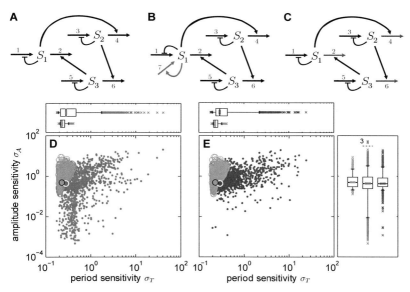

Figure 7.1.: Sensitivities of models proposed in Tsai et al. [2008] and alterations. A-C: Model structure of the models proposed in Tsai et al. [2008] and alterations: (A) negative feedback only, (B) negative feedback and positive autoregulatory feedback (marked in green), (C) negative feedback with Michaelis-Menten kinetics for degradations (marked in dark red). D: Sensitivities of the model in A (orange circles, black for median values) and of the model in B (green dots, white circle for as median). E: Sensitivities of the model in A (orange, black for median) and of the model in C (dark red dots, white for median). Box-plots in according colors characterize period sensitivity distributions (top) and amplitude sensitivity distributions (right).

feedback only (compare sections 4.3, 4.5.1, Tsai et al. [2008]). Furthermore, the same conclusions concerning the sensitivity differences between negative feedback and positive feedback models are obtained if examining the models proposed in Tsai et al. [2008] and given in Figure 7.1 A, B with the method as applied in this thesis (Figure 7.1 D). The model with negative feedback only displays higher amplitude sensitivities and lower period sensitivities than the model with positive feedback (median period sensitivities 0.23 for negative feedback only and 0.28 for positive feedback, MWU p-value $2.2 \cdot 10^{-64}$, median amplitude sensitivities 0.49 for negative feedback only and 0.42 for positive feedback, MWU p-value $6.9 \cdot 10^{-10}$, Tables A.16, A.17). However, the differences between the sensitivities of the two models are less decisive than the differences between the chain models with different feedback types (section 4.3, Figure 4.3).

Also the role of model properties other than the feedback type, i.e. matter flow and kinetics, are aimed to be examined with a sensitivity analysis for the negative feedback only model proposed by Tsai et al. [2008]. However, in this models, neither conversions nor positively regulated productions occur. Consequently, matter flow properties cannot be altered. The analysis is restricted to the examination of the influence of the kinetics employed. In the models depicted in Figure 7.1 A and B, linear kinetics (depending on $1 - S_i$) are used for productions, and regulatory Hill-terms as well as linear mass action kinetics are used for degradations. To investigate the impact of the type of kinetics, these mass action degradations in the model of panel A are replaced by saturating kinetics, i.e. by Michaelis-Menten terms depending on the respective species being degraded (marked in dark red in Figure 7.1 C). The results for the sensitivity analysis of this model with linear and with saturating kinetics are given in Figure 7.1 E. Surprisingly, applying degradations governed by Michaelis-Menten kinetics instead of linear degradation rates has a similar effect on the sensitivities as the introduction of the positive autoregulation. The period sensitivities are significantly increased while the amplitude sensitivities are significantly decreased (median period sensitivities 0.23 for the model with linear kinetics compared to 0.29 for the model with saturating kinetics, MWU p-value $2.8 \cdot 10^{-183}$, median amplitude sensitivities 0.49 for negative feedback only compared to 0.42 for saturating kinetics, MWU p-value 0.0064, Tables A.16, A.17). Indeed, the period sensitivities are even more increased if introducing saturating kinetics compared to introducing a positive autoregulation, the amplitude sensitivities are decreased less (MWU p-values $4.3 \cdot 10^{-7}$ for period sensitivity, $2.7 \cdot 10^{-5}$ for amplitude sensitivity, last two rows in Table A.17). In other words: Saturating kinetics can have a similar impact on the sensitivities of the period and amplitude as a positive feedback. Thus, it is not only the feedback which needs to be considered, but especially also the types of kinetics applied in order to characterize the sensitivities of the oscillatory properties.

Contrary to the chain models, in the negative feedback model proposed by Tsai et al. [2008], applying saturating kinetics decreases amplitude sensitivity instead of increasing it. The reason for this could lie in the fact that in the latter model, the species obey conservation relations. Consequently, the maximal concentration and thus also the amplitude of the three species is limited to one. If saturation occurs, the amplitude might yield more frequently maximal or close to maximal values thereby limiting the alterations which could occur. Indeed, for the model with positive feedback (Figure 7.1 B), there is a significant negative correlation between the amplitude and the amplitude sensitivity ($R_S = -0.45$). For the model with saturating kinetics, however, there is a lack of such negative correlation ($R_S = 0.094$), even if considering single amplitudes and the according single amplitude sensitivities ($R_S = -0.042$ for the amplitude of S_1 and its sensitivity, $R_S = -0.033$ for S_2 and $R_S = -0.055$ for S_3). Consequently, the origin of the decrease in amplitude sensitivity for saturating kinetics is a different one than for the positive feedback, and it necessitates further examination to fully characterize it.

7.2. Inferring structural properties from sensitivities

Previously, it has been shown that positive feedback (chapter 4), saturating kinetics (chapter 6) as well as lacking matter flow between species (chapter 5) can lead to increased period sensitivities. Positive feedback decreases amplitude sensitivities. The analyses of the effects of certain model properties on period and amplitude sensitivities has been performed with the help of prototype oscillator models, the chain models. In this part, it is investigated to which extent the obtained insights of chapters 4 - 6 remain valid for models of cellular processes. For this purpose, only feedbacks and kinetics will be considered since matter flow properties only slightly influence the sensitivities and can thus hardly be inferred from sensitivities.

For the examination, it is focused on models for circadian and calcium oscillations (compare chapter 3). Calcium and circadian rhythms differ in the function of their oscillations and consequently, the systems should be designed differently to fulfill their purposes. Circadian oscillations need to supply a timing information, and hence the period should be very stable if the system suffers (external) disturbances. To render this possible, the feedback structure and kinetics should enable low period sensitivities. In contrast, the period of the calcium oscillations are discussed to encode signals and needs to be easily adaptable to environmental changes. Therefore, calcium oscillation models are allowed and designed to exhibit high period sensitivities.

Three models of circadian oscillations and three models of calcium oscillations are examined, one of each is the model analyzed in the beginning of this work (chapter 3). For circadian oscillations, models for different organisms are inspected: (i) the model for mammalian cells examined in chapter 3 (Becker-Weimann et al. [2004]), (ii) a model established for fruit flies (*Drosophila*, Goldbeter [1995]) and (iii) a model for plants (*Arabidopsis thaliana*, Locke et al. [2005b]). For calcium oscillations, the analysis is restricted to IP_3-mediated intracellular calcium oscillations: besides (i) the phenomenological model examined in chapter 3 (Goldbeter et al. [1990]), (ii) an open-cell model (Sneyd et al. [2004]) and (iii) a closed-cell model (De Young and Keizer [1992]) reflecting the two major types of calcium models are analyzed. A detailed description of the models, the ODEs and the reference parameter set published together with the models are given in the Appendix, section A.7.1.

The examination is divided into three steps: First, perform a sensitivity analysis of the model as described in section 2.3. Second, compare the obtained sensitivities to those of the chain models and derive a prediction of the model structure. Third, examine whether the actual model structure coincides with the predicted structure, and if not, try to further investigate the reason.

7.2.1. Results of the sensitivity analyses

The period sensitivities and amplitude sensitivities obtained for the six chosen models are given in Figure 7.2 and Table A.18 in the Appendix. The distributions among the circadian rhythm models, among the calcium oscillations models and between the circadian and calcium oscillations models are compared. It is focused on the medians, the dispersion and the sensitivities at the respective reference parameter, that is the parameter set published together with the model.

First, the circadian oscillation models are compared (Figure 7.2 A-C, upper three box-plots in G, three box-plots on the left in H). The mammalian circadian rhythm model of Becker-Weimann and colleagues [2004] is the only one which displays low period sensitivities for all parameter sets (Figure 7.2 A). Such a pattern one would

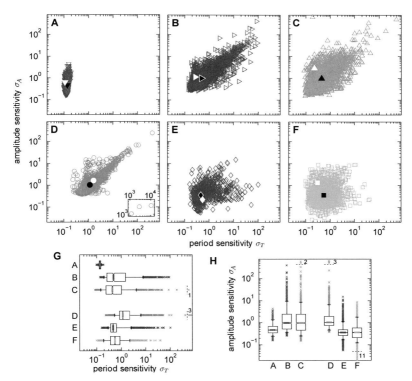

Figure 7.2.: Sensitivities for different models of circadian and calcium oscillations. A-C: Circadian models for different species. (A) mammals (red, triangle edge down, 613000 sets examined, Becker-Weimann et al. [2004]); (B) Drosophila (dark red, triangle edge up, 173000 sets examined, Goldbeter [1995]); (C) Arabidopsis (orange, triangle edge right, 364000 sets examined, Locke et al. [2005b]). D-F: Calcium models. (D) phenomenological model (blue, circle, 71400 sets examined, Goldbeter et al. [1990]); (E) open-cell model (dark blue, star, 58700 sets examined, Sneyd et al. [2004]); (F) closed-cell model (light blue, rectangle, 155000 sets examined, De Young and Keizer [1992]), number of sets examined until 2500 parameter sets yielding regular oscillations found. Black symbols denote median values, white symbols the sensitivity for the reference parameter set, i.e. the parameter set given in the original literature. G, H: Box-plots of the period (G) and the amplitude (H) sensitivity distributions for the models from A-F.

expect if considering that the function of the period in circadian rhythm is to remain stable facing environmental changes. Contrarily, the models for *Arabidopsis* by Locke and colleagues [2005] and *Drosophila* by Goldbeter [1995] exhibit larger median period robustness and decisively larger variances of the period sensitivity (in this order, Figure 7.2 C, B, G, Table A.18, MWU p-values $< 10^{-10}$, Table A.19). Concerning the amplitude sensitivities, the same order applies, the mammalian model displays the lowest amplitude sensitivity and dispersion of the amplitude sensitivity, followed by that of the plant and the fruit fly (Figure 7.2 H, MWU p-values $< 10^{-3}$, Table A.19). The reference parameter sets published together with the circadian models always display a rather low period sensitivity (compared to what is possible with the same model and other parameters) and an amplitude sensitivity only slightly above median values (white and black symbols in Figure 7.2 A-C).

For the calcium models, only the phenomenological model published by Goldbeter and colleagues [1990] (Figure 7.2 D) displays decisively higher period sensitivities than obtained for the circadian rhythm models (Figure 7.2 G). Indeed, the period sensitivity distributions of the open cell calcium model proposed by Sneyd and colleagues [2004] (Figure 7.2 E) and the closed cell calcium oscillations model published by DeYoung and colleagues [1992] (Figure 7.2 F) do not differ significantly from that obtained for the *Drosophila* circadian oscillations model (Figure 7.2 G, lines 11 and 12 in Table A.19 with high-lighted MWU p-values ≥ 0.13). The amplitude sensitivities of the phenomenological model (Figure 7.2 D) are larger than those of any of the other five models (Figure 7.2 H). In contrast, the other two calcium oscillations models yield the lowest amplitude sensitivities among all six models (Figure 7.2 H), and these two do not significantly differ (MWU p-value 0.089, Table A.19). For the reference parameter sets (white symbols in Figure 7.2 D-F), no regularity of the sensitivities compared to the median values (black symbols in Figure 7.2 D-F) as for the circadian rhythm models can be identified, both smaller and larger sensitivities are obtained.

7.2.2. Circadian rhythm models

The circadian rhythm model proposed in Becker-Weimann et al. [2004] displays particularly low period as well as amplitude sensitivities (7.2 A). From the sensitivities, one would predict that the model yields only negative feedback loops and, except for the regulatory interactions, exclusively linear kinetics. This prediction is further strengthened by the dispersion of the sensitivity distributions being small (Figure 7.2 G, H) which is observed only for the chain models with negative feedback and linear kinetics (sections 4.3, 5.3).

A visualization of the structure of the model is given in Figure 7.3 A. Indeed, except for the regulated productions ν_1, ν_3 and ν_8, all reactions exhibit mass action kinetics favoring low sensitivities. The feedbacks of the model are detected by examining the Jacobian matrix at the reference parameter set with a self-implemented

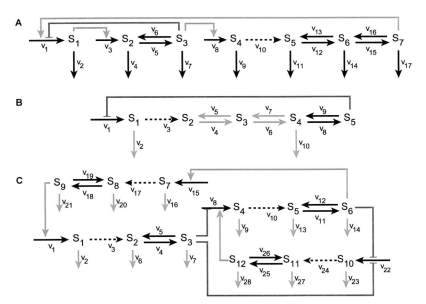

Figure 7.3.: Structure of the examined circadian oscillations models. A: Mammalian circadian rhythm model proposed by Becker-Weimann et al. [2004]. B: *Drosophila* circadian rhythm model proposed by Goldbeter [1995]. C: *Arabidopsis* circadian oscillations models proposed by Locke et al. [2005b]. ODE systems are given in the Appendix, section A.7.1. Activating regulations are marked by green arrows, inhibiting regulations by red arcs. Reactions employing saturating kinetics for a source species are marked by blue arrows.

Matlab (Matlab 7.11, The MathWorks, Inc.) function (source code available on request). The examination reveals that not only negative feedback loops, but also positive feedback loops are present in the mammalian circadian oscillations model. It yields one positive feedback loop of length seven, one negative feedback loop of length three, three positive feedback loops of length two, and seven negative feedback loops of length one. The negative feedback loop of length three is the transcription-translation feedback loop being a characteristics of each circadian oscillation (compare section 3.1.1). The feedback loops of length one, the autoregulatory feedback loops, are due to all species being source species of at least one reaction and do not need further examinations. Although there are four positive feedback loops present, they do not strongly increase the period sensitivities of the model. In fact, the positive feedback loop of length seven is set in row and acts outside the negative feedback loop. This is exactly the same situation as for the negative feedback chain model in which also a positive feedback loop is wrapped around the negative (chapter 4). It results in a lack of influence of the positive feedback loop. The positive feedback loops of length two originate from the reversibility of the linear transport and activation processes encoded in reactions ν_5, ν_6, ν_{12}, ν_{13} and ν_{15}, ν_{16}. They contain no non-linearities and thus may be too weak to disturb the overall sensitivity. Therefore, the weak positive feedbacks do not noticeably affect the sensitivities, and still low sensitivities are observed in the mammalian model of circadian oscillations. Thus, for determining the effect of positive feedbacks on the sensitivity, also the location of the feedback loop in comparison to the negative feedback loops of the systems as well as the strength of the feedbacks need to be considered.

The *Drosophila* model of circadian oscillations (Goldbeter [1995], Figure 7.2 B) displays a period sensitivity distribution slightly decreased compared to that of the negative feedback chain model with saturating kinetics (section 6.3.2). Its amplitude sensitivities resemble rather that of the positive feedback chain model with saturating kinetics (section 6.3.2). However, the positive feedback chain model delivers decisively larger period sensitivities and the negative feedback chain model only slightly higher amplitude sensitivities. Consequently, one would predict that the model relies on negative feedback and employs mainly saturating kinetics.

The *Drosophila* circadian rhythm model published by Goldbeter [1995] is visualized in Figure 7.3 B. Indeed, six out of the ten reactions yield Michaelis-Menten kinetics (reactions marked in light blue in Figure 7.3 B), which can be saturated (saturation index > 0.8) or not, depending on the values of the parameter set. The influence of the saturation level on the sensitivity is examined in the following. It can be observed that the median period sensitivity as well as amplitude sensitivity is increasing with increasing number of those reactions being saturated (Figure 7.4 A-C). Considering only the median values, however, the correlation is only significant at level 0.05 but not at level 0.01 due to the small number of pairs which are compared (Table 7.1, upper two lines). In contrast, there is a significant positive correlation between the number of saturated parameter sets and the sensitivities if comparing all 2500 parameter sets separately (Table 7.1, third line). Thus, as

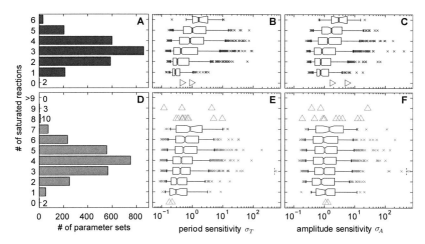

Figure 7.4.: Saturation effects for the *Drosophila* and *Arabidopsis* circadian oscillations models. In each model, a reaction ν of one of the six or 12, respectively, reactions employing Michaelis-Menten kinetics is considered to be saturated whenever its general saturation index satisfies $sat_\nu > 0.8$. A and D: Number of parameter sets obtained for each number of saturated reactions for the *Drosophila* model from Goldbeter [1995] (A) and the *Arabidopsis* model from Locke et al. [2005b] (D). Numbers sum up to 2500 for both panels. B and C: Period (B) and amplitude (C) sensitivity distributions for each number of saturated reactions for the *Drosophila* circadian rhythm model. E and F: Period (E) and amplitude (F) sensitivity distributions for each number of saturated reactions for the *Arabidopsis* circadian rhythm model.

Table 7.1.: Correlations between saturation level and sensitivities measured by Spearman's rank correlation R_S for the circadian rhythm models in *Drosophila* and *Arabidopsis*. Given are the correlations between the number of saturated reactions (# sat rct) and the according median sensitivity (μ) for all numbers of saturated reactions ("all") or only for those where more than ten data points are available ("> 10"; $1-6$ and $1-7$ for the *Drosophila* and *Arabidopsis* model, respectively) or the sensitivities of all parameter sets (σ), as well as the according p-values (compare section 2.4.5). Data corresponds to Figure 7.4.

	correlation between		*Drosophila* model				*Arabidopsis* model			
			period	p-val	ampl	p-val	period	p-val	ampl	p-val
all	# sat rct vs.	μ	0.79	0.027	0.25	0.27	0.84	0.0059	-0.067	0.42
>10	# sat rct vs.	μ	1	0.013	1	0.013	1	0.0072	0.21	0.30
all	# sat rct vs.	σ	0.35	$1.7 \cdot 10^{-67}$	0.22	$4.3 \cdot 10^{-29}$	0.17	$4.3 \cdot 10^{-17}$	0.044	0.014

predicted, the saturating kinetics lead to increased sensitivities.

The feedback structure is analyzed with the help of the Jacobian matrix at the reference parameter set. Additionally to the obvious negative feedback loop of length five, five negative feedback loops of length one (one autoregulatory feedback loop for each species in the model), as well as three positive feedback loops of length two are revealed. These positive feedback loops are due to reversible modification and transport by reactions ν_4 - ν_9. ν_8 and ν_9 employ linear kinetics, and thus the positive feedback from S_4 to S_5 and back may be too weak to noticeably affect the sensitivities. This would be similar to the mammalian circadian oscillations model in which only low period sensitivities are observed despite a positive feedback loop of length two. In contrast to reactions ν_8 and ν_9, reactions ν_4 - ν_7 are modeled using saturating kinetics and thus the positive feedback established by the non-linear conversions cycles may considerably contribute to the observed increase in sensitivity. To which extent these non-linear, but short positive feedback loops originating from reversible processes influence the sensitivities remains to be investigated in a prospective work.

The *Arabidopsis* circadian rhythm model (Locke and colleagues [2005b]) displays similar period and amplitude sensitivities as the *Drosophila* model (compare Figure 7.2 C, G, H), and thus also the prediction for its underlying structure from the sensitivity is similar. The model is predicted to employ very frequently saturating kinetics which also constitutes the source of increased sensitivities, and most probably rather negative instead of positive feedback.

The structure of the *Arabidopsis* circadian rhythm model is visualized in Figure 7.3 C. In fact, each of the 12 species from the model is degraded by reactions following Michaelis-Menten kinetics (marked in blue in Figure 7.3 C) thus perfectly matching the prediction. In fact, the period sensitivity significantly increases with increasing number of saturated reactions, but there is no significant correlation between the amplitude sensitivity and saturation (Figure 7.4 D-F, right column of Table 7.1). This could explain the observed low amplitude sensitivities compared to the negative feedback chain model with saturating kinetics (section 6.3.2). However, the reason why the saturation only affects the period sensitivity here is not resolved yet.

Additionally, the analysis of the feedback structure with the help of the Jacobian matrix at the reference parameter set reveals that positive and negative feedback loops occur in the model. There are three negative feedback loops of length twelve, nine or six, twelve negative feedback loops of length one (one autoregulatory feedback loop for each species) and four positive feedback loops of length two. These positive feedback loops arise due to the reversible transport of each protein between cytosol and nucleus. They consist only of linear reactions, and since these linear positive feedback loops revealed to lack any noticeable impact on the period sensitivities in the mammalian circadian model, one could conclude that they do not remarkably affect the sensitivities in this model either. In fact, in this model, the transcription-translation feedback loop being typical for circadian oscillations is the negative feedback loop of length nine. The two additional negative feedback loops could be important for the reduced impact on the effect of saturation on the amplitude sensitivity compared to the negative feedback chain model with saturating kinetics yielding only one non-autoregulatory negative feedback loop (section 6.3.2).

7.2.3. Calcium oscillations models

The phenomenological calcium oscillations model proposed by Goldbeter et al. [1990] yields very high period and amplitude sensitivities (Figure 7.2 D) compared to all six examined models. Its period sensitivity distribution lies between that of the positive feedback chain model lacking matter flow (section 5.3) and that of the positive feedback chain model with saturating kinetics (section 6.3.2); its amplitude sensitivity distribution lies between that of the positive feedback chain model with saturating kinetics and that of the negative feedback chain model with saturating kinetics (both section 6.3.2). Consequently, the model is predicted to rely on positive feedbacks and to employ saturating kinetics.

In fact, the model yields a strongly connected structure: Its only two variables are

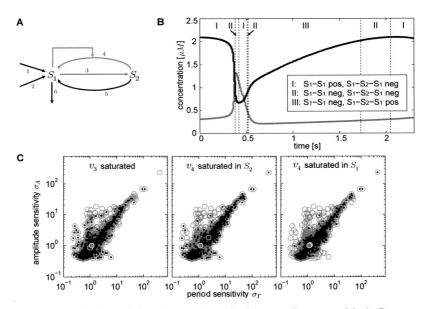

Figure 7.5.: Analysis of the phenomenological calcium oscillations model. A: Structure of the phenomenological calcium oscillations model proposed in Goldbeter et al. [1990]. B: Concentration changes of S_1 (blue) and S_2 (black) and the according feedback loop structure of the model over one limit cycle for the reference parameter set. The autoregulatory feedback loop of S_2 is always negative, the autoregulatory feedback loop of S_1 (S_1-S_1) can change from positive (pos, I) to negative (neg, II) and back. The feedback loop comprising both species (S_1-S_2-S_1) can switch between negative (I and II) and positive (III). C: Sensitivities of all 2500 parameter sets (blue circles, white bordered circle for median values) compared to sensitivities for those parameter sets with ν_3 being saturated (in S_1), ν_4 being saturated in its source species S_2, ν_4 being saturated in the regulation by S_1 (black dots, black circle for median values).

connected by three reactions, and three more reactions only affect one of the species (model visualized in Figure 7.5 A). At the unstable steady state of the reference parameter set, according to the Jacobian matrix, there is a positive autoregulatory feedback loop of S_1, a negative autoregulatory feedback loop of S_2 and a negative feedback loop comprising both species. Due to the interfering regulations, in the course of an oscillation, the feedback loops present in the system change their type (Figure 7.5 B). In particular, the feedback loop of S_1 on itself turns several times from positive to negative and back (changing from I to II or back in Figure 7.5 B); and if the autoregulatory feedback loop on S_1 is negative, the feedback loop of length two turns positive and negative once (changing from II to III and back in Figure 7.5 B). This is caused by S_1 having different effects on itself as well as on S_2. Via the degradation reaction ν_6, S_1 has a negative effect on itself, but via reaction ν_4, it has a positive effect on itself (Figure 7.5 A). Additionally, via ν_4, S_1 acts negatively on S_2, via ν_3, it acts positively on S_2. In contrast S_2 acts only negatively on itself and positively on S_1 (both via reactions ν_4 and ν_5, Figure 7.5 A). Consequently, two ambivalent feedback loops occur in which the feedback type can switch between positive and negative during an oscillation: one autoregulatory on S_1, the other comprising S_1 and S_2 (Figure 7.5 B).

Additionally, in contrast to the circadian models in which regulatory Hill-like terms occur only in regulated productions, in this model, two of the conversions also yield activating Hill-terms (reactions ν_3 and ν_4, marked in blue in Figure 7.5 A). Their reaction rates thus have the potential to saturate and can contribute to the increased sensitivities. However, the sensitivity does not increase monotonously with an increasing number of saturated reactions. Instead, the saturation of the different reaction has different effects on the sensitivities (Figure 7.5 C). Saturation of species S_1 in ν_3 slightly decreases both period and amplitude sensitivities, but saturation of S_2 in reaction ν_4 strongly increases both sensitivities. The saturation of the regulatory term in ν_4 depending on S_1 slightly increases the sensitivities. Thus, as predicted, also saturation plays a role for the sensitivities in the phenomenological calcium model (Goldbeter et al. [1990]).

The open cell calcium oscillations model as proposed by Sneyd and colleagues [2004] displays very low amplitude sensitivities (Figure 7.2 E), the median is even lower than for the positive feedback chain model which yields the lowest median amplitude sensitivity observed for all chain models (section 4.3). One would thus predict that positive feedback may be dominant in the system. However, the period sensitivities are not as low as for the negative feedback chain model (section 4.3), but also not as high as for the positive feedback chain model. This rather indicates that only negative feedbacks are present with very few reactions governed by saturating kinetics (compare section 6.3.2). Thus, the information regarding the combination of period and amplitude sensitivities are not consistent, and it is difficult to conclude on structural properties. One could either predict that only negative feedback with few reactions yielding saturating kinetics occur, or that positive feedbacks are dominant in the system.

The model structure is given in Figure 7.6 A. Analysis of the feedbacks at the unstable steady state of the reference parameter set reveals the presence of 30 feedback loops: 21 negative and nine positive feedback loops. The longest feedback loops are two of length five, one negative, one positive, and five negative and two positive feedback loops of length four. Except for S_2, each species possesses a negative autoregulatory feedback loop. In the course of a limit cycle oscillation with the parameters from the reference parameter set, five of the positive feedback loops become negative. In detail, this happens because the interaction $S_3 \rightarrow S_4$ turns from positive to negative once (and back). In other words, the model is highly connected. The occurrence of a high number of feedback loops could explain why the predictions from the sensitivities are ambiguous and neither the sensitivity results of the negative feedback chain model nor of the positive feedback chain model can be applied. The most connecting species is S_4 participating in 19 of the 24 non-autoregulatory feedback loops. S_1 occurs in 17 out of the 24 reactions in the model and participates in 18 of the 24 non-autoregulatory feedback loops.

Concerning the kinetics, the open cell calcium model by Sneyd et al. [2004] displays frequently saturating kinetics, but exclusively with respect to species S_1. 15 of the 24 reactions of the model can saturate in S_1. However, among these 15 different reactions only eight different possibly saturating terms in S_1 occur (Figure 7.6 B) which are multiplied by different combinations of species or nl-parameters in different reaction rates. Considering their saturation separately, only four of the eight kinetics terms alter the sensitivity significantly: Saturation in reaction ν_3 reduces the amplitude sensitivity, saturation in reactions ν_{16} and ν_{22} or reactions ν_{10}, ν_{13}

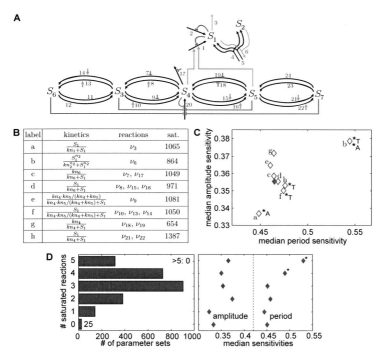

Figure 7.6.: Analysis of the open cell calcium oscillations model. A: Structure of the open cell model proposed by Sneyd et al. [2004]. Green arrows: activating regulations, red arcs: inhibiting regulations, blue arrows: reactions employing saturating kinetics for a source species, green and red arrowheads: regulation by species S_1. B: Possibly saturating kinetics terms occurring in the model, and in which reactions (2nd column). Column "sat.": Numbers of parameter sets for which this term is saturated. C: Median sensitivities for the terms from B being saturated according to their label (empty diamonds), the median value for all 2500 sets is given by the filled diamond. Significant differences (MWU p-value < 0.01) of the sensitivity distributions of the parameter sets if the term is saturated (diamond) compared to the term being not saturated (not shown) are marked by $^*T$ or $^*A$ for period sensitivity or amplitude sensitivity, respectively. D: Numbers of parameter sets with zero to five of the reaction terms from B being saturated (left). The according median sensitivities for each category are displayed to the right separated by the dotted line. Stars indicate significant differences.

and ν_{14} increases the period sensitivity and saturation in reaction ν_6 increases both period sensitivities and amplitude sensitivities (Figure 7.6 C). Counting the overall number of saturated reactions, only the period sensitivity displays a significant increase for four or more terms being saturated (Figure 7.6 D). Thus, kinetics play a role for the sensitivities, but the influence of the feedback loops seems to be more decisive in the open cell calcium oscillations model (Sneyd et al. [2004]).

The closed cell calcium oscillations model proposed by DeYoung and colleagues [1992] displays sensitivities resembling those of the open cell calcium oscillations model. Low amplitude sensitivities and intermediate period sensitivities are obtained (Figure 7.2 F, G, H). Thus, again, from the insights of the chain models, positive feedback could be predicted as well as negative feedback with few reactions yielding saturating kinetics.

The model structure is visualized in Figure 7.7 A. An analysis of the feedbacks on the limit cycle for the reference parameter set reveals even more feedback loops than for the open cell calcium oscillations model: In the closed cell calcium oscillations model, there are 189 feedback loops in total. Of these, nine are negative autoregulatory loops (one for each species), ten negative feedback loops of length nine, and 55 are negative feedback loops of lengths three to eight, 115 are positive feedback

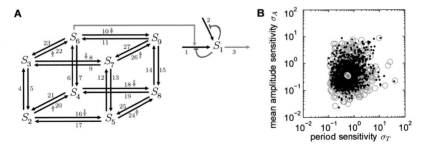

Figure 7.7.: Analysis of the closed cell calcium oscillations model. A: Structure of the closed cell model proposed by De Young and Keizer [1992]. Green arrows: activating regulations, red arcs: inhibiting regulations, blue arrows: reactions employing saturating kinetics for a source species, green arrowheads: positive regulation by species S_1. B: Sensitivities for parameter sets with ν_3 being not saturated (blue, 1045 parameter sets) or with ν_3 being saturated (black, 1455 parameter sets).

loops (eight of which of maximal length nine). S_1 occurs in 113 of the 189 feedback loops, but species S_6 is even more often present, in 158 of 189 feedback loops. Thus, the model is highly connected, and predictions cannot be made on the basis of the sensitivity analyses of the less connected chain models.

Concerning the kinetics, the closed cell model only yields one reaction with potentially saturating kinetics, which is reaction ν_3. Its saturation yields a slight but significant decrease in the period sensitivities (median values 0.58 for ν_3 being non-saturated versus 0.53 for ν_3 being saturated, MWU p-value: 0.0018), but no significant change for the amplitude sensitivities (median values 0.34 and 0.36 for ν_3 being not saturated and saturated, respectively, MWU p-value: 0.029, Figure 7.7 B). Thus, one can conclude that also here the feedback loops are more decisive for the sensitivities than the kinetics.

7.3. Summary, discussion and outlook: Further applications

In this chapter, the results found for the sensitivities of the chain models in chapters 4, 5, 6 are applied to other model types. First, the core oscillator models with negative feedback only and with negative feedback and an autoregulatory positive feedback loop proposed in Tsai et al. [2008] are examined with the help of the sensitivity analysis introduced in this thesis. The results of the sensitivity analysis are in accordance with the results from the publication in which it has been argued on the operational range of the period and amplitude with respect to changing one particularly chosen parameter: negative feedback only leads to low period sensitivities (or low tunability) of the period whereas an additional positive feedback loop increases the period sensitivities and yields lower amplitude sensitivities.

Additionally, for the negative feedback only oscillator model employed in Tsai et al. [2008] it is shown here that the kinetics play a decisive role for the period and amplitude sensitivities. In particular, it is found that not only for the chain model saturating kinetics can increase the period sensitivities obtained, but also for other types of models (section 7.1). Therefore, this thesis provides an extension of the results shown in Tsai et al. [2008] where the importance of the feedback type on sensitivity has been high-lighted. However, in contrast to the chain models, for the model proposed by Tsai et al. [2008], saturating kinetics *reduce* the observed amplitude sensitivities and thus act similarly to an additional positive feedback. This might be due to conservation conditions hidden in the model of Tsai and colleagues [2008], which has been derived from conversion cycles and thus the maximal amplitude of each of the modeled species is constrained. The effect of conservation relations of the model species on the sensitivities of the oscillatory properties has not been investigated in this thesis, but it would be possible to tackle also this issue in the context of an altered chain model. It would be interesting to pursue in this direction as a prospective work.

Furthermore, in this chapter the sensitivities of different models of circadian and calcium oscillations are investigated and related to their model structure (section 7.2). In general, for models yielding few feedbacks and thus resembling the chain models, predictions of the structure from their sensitivities on the basis of the results for the chain models are appropriate. This is the case for the herein examined circadian rhythm models. In particular, the finding that saturating kinetics lead to larger sensitivities for models lacking conservation relations can be applied for deriving predictions on the prevailing kinetics. Furthermore, as an additional extension to the work of Tsai and colleagues [2008], it is shown that not only the existence of positive feedback loops, but also their position and strength determine whether they yield an influence on the sensitivities or not. In detail, positive feedback loops outside of the negative feedback loop (which is essential for oscillations and thus can be found in those models) as in the chain model with negative feedback and in the mammalian circadian model proposed by Becker-Weimann and colleagues [2004] are affecting the sensitivities only slightly. Additionally, for short positive feedback loops resulting from reversible processes which are modeled by two reactions in opposite directions between two species thus resembling a conversion cycle, the kinetics might be of importance for them to be influential on the period and amplitude sensitivities. Linear conversion cycles seem not to increase the sensitivities (at least in the mammalian circadian model), conversion cycles with saturating kinetics might be more influential (section 7.2.2). The detailed impact of such motifs remains to be investigated as part of a prospective work.

As soon as models become more connected and differ more strongly from the classical chain model structure, predictions on the structure from the sensitivities based on observations made for the chain models could not be drawn. Extreme cases of highly connected models are the open and closed cell calcium oscillations models (Sneyd et al. [2004], De Young and Keizer [1992]), which contain 24 and 180 non-autoregulatory feedback loops, respectively. Additionally, in contrast to the chain models, these two models employ conservation relations due to modeling probabilities in five or even eight variables, respectively (i.e. it is $\sum_{i=3}^{7} S_i < 1$ for the open cell model and $\sum_{i=2}^{9} S_i = 1$ for the closed cell model). In accordance with findings of decreased amplitude sensitivities for saturating kinetics in the model proposed by Tsai and colleagues [2008] (section 7.1), this fact might also play a role for the extremely low amplitude sensitivities encountered in these two models.

7.3.1. Applicability of structure prediction

In fact, the prediction of the model structure from the results of a detailed sensitivity analysis is an artificial situation: As the sensitivity analysis is performed using a model of the processes, of course the structure of the model and thus of the underlying process is known. Therefore, as part of this thesis, section 7.2 is rather to be seen as testing to which extent the results found for the sensitivities of the chain models can also be employed for models of biological processes.

To really determine the network structure of a process from the sensitivity results obtained from observations of experiments, one needs to ascertain that the perturbations applied to the system are similar to those employed for determining the sensitivity. In this case, small perturbations of single parameters are applied. As soon as the system is subject to stronger perturbations in the experiments, the rate with which high sensitivities have been observed might increase - or even oscillations may cease. Both coincides with passing bifurcations, either one of the HBs such that the oscillating regime is left, or other bifurcations which keep the oscillatory behavior but may decisively change the period or amplitude. These bifurcations occur already in systems as simple as the positive feedback chain model (section 4.4.3).

Additionally, except for the period sensitivities of the negative feedback chain model and positive feedback chain model with both linear kinetics, the sensitivity distributions of the models overlap, and they are rather distinguished in terms of their median values. In an experimental setting, one would consider most frequently only few conditions, few cell types or few nutritional media sources, which correspond to few, rather similar parameter sets of the underlying process. One could increase the number of observed "parameter sets" by also investigating other cell types, or other species, but then one cannot exclude that also structural properties are altered together with the kinetic parameters. In other words: In experiments, the whole potential of the structure of the model system is difficult to be detected, and one even does not know whether the observed sensitivities are rather found in the center or at the lower or upper border of the sensitivity distribution being possible with the system. Thus, even for structurally very different systems, the experimentally observed sensitivities might be similar and conclusions on different structural properties are difficult to draw from them.

7.3.2. Suggesting structure to tune sensitivity

In addition to the possible application discussed so far, the results obtained in this thesis could also be employed to adapt the sensitivities as desired. To achieve this, for a system of which the structure is known, the required changes and alterations can be suggested on the basis of the findings made in this work. Most appropriate for this application is the field of synthetic oscillators (reviewed in e.g. Purcell et al. [2010]). Synthetic oscillators are designed to display oscillatory behavior in their components. One of the earliest synthetic oscillators is the so-called "repressilator", which has been instantiated in *Escherichia coli* (Elowitz and Leibler [2000]). It is based on three genes expressing one protein each which represses the transcription of one of the other genes. Thus, a negative feedback loop is established. As one would predict from the results of the chain models, a sensitivity analysis of the model of the repressilator reveals very low period sensitivities (Appendix, section A.5.1, Figure A.26 A). By introducing saturating kinetics in the system, the period sensitivities can be decisively increased as is indicated by an according sensitivity analysis (Appendix, Figure A.26 A, Tables A.20, A.21).

A similar behavior is observed for the model of another synthetic oscillator, the "metabolator", which combines genetic regulation and metabolism and has been also realized in *Escherichia coli* (Fung et al. [2005]). A sensitivity analysis shows that the metabolator model already displays larger sensitivities than the repressilator, as it yields additional feedback loops and saturating kinetics. However, at least the amplitude sensitivities are significantly further increased if applying more reactions with saturating kinetics (Appendix, Figure A.26 B, Tables A.20, A.21).

A similar increase in period sensitivities, but a decrease in amplitude sensitivities might be observed for introducing a positive feedback. Whether the system indeed responds as expected remains to be investigated in each particular case, but positive feedback and saturating kinetics are means to increase period sensitivities in established oscillators, and maybe even to decrease amplitude sensitivities, especially in the case of components of the system being available only in limiting amounts.

7.3.3. Determining sensitivity to obtain indications on the biological potential of oscillations

Apart from conclusions on the model structure from sensitivity or from sensitivity on the model structure, the mere results of a sensitivity analysis as suggested in this thesis could yield valuable information for a model of a process in question. Thereby, one could detect common characteristics of parameter sets with very low period sensitivities (if present). If the system yielded these characteristics, it could be enabled to act as time-keeper. Similarly, conditions prevailing for parameter sets with high period or amplitude sensitivities (if present for the system) could be derived. High period or amplitude sensitivities can hint on the ability to encode signals in the according oscillatory property. Thus, under the conditions detected for parameter sets with high sensitivities, the system might be able to act in frequency or amplitude encoded signaling.

There are several biological systems in which oscillations are observed whose sensitivity potential is not yet explored. One example is the canonical NF-κB signaling in which sustained NF-κB-oscillations have been observed on single cell level whose period has been shown to affect their target gene expression (e.g. Nelson et al. [2004], Sung et al. [2009], Ashall et al. [2009]).

An analysis of the sensitivity of the oscillatory properties of the process via the sensitivity analysis of a reduced NF-κB model (original model proposed by Kearns et al. [2006], reduction: J. Mothes, personal communication, to be published, ODE system in the Appendix, section A.7.8) shows that both very low and very high period sensitivities are possible (see the Appendix, section A.5.2). For single parameter sets, the sensitivity of the NF-κB model is very low ($\sigma_T \leq 0.1$) thus hinting to the possibility that the oscillations could potentially yield a timing information with stable period (a clock function for NF-κB oscillations has been also suggested in Paszek et al. [2010]). For other parameter sets, the period sensitivity is very high ($\sigma_T \geq 10$) thus hinting to the potential of encoding a signaling information in the variable period as well. A detailed examination of the parameter sets at the extremes of the period sensitivity distribution could deliver possible conditions under which the period of NF-κB oscillations is especially sensitive or robust. Similar considerations are possible for the amplitude sensitivities.

Note that the definition of the sensitivity in this work includes small variation of all kinetic parameters of the models. Especially in the case of signal transduction, only few parameters might be varied more strongly according to a given stimulus. Hence, even parameter sets responding weakly on average to small changes in all parameters might be strongly responsive to a strong change in only one or few parameters. Still, the sensitivities σ_T and σ_A obtained by the sensitivity analysis in this thesis can provide an indication on the actual responsiveness of the period and amplitude towards stimuli and could be employed as tool to better understand the biological potential of oscillations.

8. Discussion: Mechanisms underlying the robustness of oscillatory properties

In this thesis, the sensitivities of the oscillatory properties period and amplitude towards perturbation in parameters are examined. It is investigated which model structures such as feedback type, matter flow properties or kinetics lead to low or high period or amplitude sensitivities. For this examination, a chain model as prototype oscillator model is used, and the sensitivities σ_T and σ_A are calculated as the relative change of the period for a relative change in the parameter, averaged over all parameters.

Using this setting, it is found that the model structure is highly important for the potential period and amplitude sensitivities which a model, and thus its underlying biological process, exhibits. In accordance with results obtained for examining the tunability of period and amplitude for changes in one particular parameter (Tsai et al. [2008], Stricker et al. [2008]), a single negative feedback model is shown to exhibit oscillations with low period sensitivities but variable amplitude sensitivities, whereas a substrate-depletion oscillator relying on a single positive feedback leads to oscillations with comparably higher period sensitivities, but lower amplitude sensitivities (chapter 4). In addition, in this work the effects of matter flow properties between species on period and amplitude sensitivity are addressed. Changing matter flow properties can alter the sensitivity of the period or amplitude towards perturbations of particular reactions (section 5.4.1). In systems lacking matter flow between species, compensatory regulations (section 5.4.3) or even regulations enabling the occurrence of oscillations (section 5.2.1) may be disrupted leading to increased sensitivities or the lack of sustained oscillations. Besides, in this thesis, the differences in period and amplitude sensitivities for using saturating kinetics compared to linear kinetics are examined. As a first step, for steady states, it is found that not only in conversion cycles, but also already the steady state of a species being subject to a possibly saturating degradation can exhibit ultrasensitivity (section 6.2). Interestingly, the results from the steady state sensitivity also apply for the sensitivities of oscillatory characteristics. The period and amplitude sensitivities of both chain models increase with increasing number of saturated reactions (section 6.3.4). Thereby, the saturations of different reactions reveal differently strong impacts (section 6.3.5).

Opposed to steady states which are mainly characterized by the value and their stability, oscillations have a variety of different properties whose sensitivity or robustness can be examined. For example, besides the period and amplitude, also the phase constitutes a property which can be important in the biological process (examined e.g. in Bagheri et al. [2007]), as well as entrainability of oscillations which is a crucial feature of circadian oscillations (Leloup and Goldbeter [2004], Bagheri et al. [2008], Hogenesch and Ueda [2011]). Furthermore, most frequently in literature, robustness of oscillations is associated with oscillation probability (e.g. Ma and Iglesias [2002], Wagner [2005], Wong et al. [2007], Tian et al. [2009], Xu and Qu [2012]).

Since multiple properties of the oscillator might be important concerning its embedding into a biological process, robustness trade-offs are interesting to examine. Robustness trade-off means that not all oscillatory properties might be similarly robust, but some might be robust whereas others are sensitive. This implies that a "law of conservation of fragility" may account (Csete and Doyle [2002]), and the structure of the system underlying the oscillations has to adapt in order to find a compromise between the system features and their robustness. Trade-offs in robustness have been discussed to be usually present in complex systems (Csete and Doyle [2002], Stelling et al. [2004b], Kitano [2007]). A robustness trade-off between different system features has been observed in an analysis using ODE models (Wong et al. [2012]), but another analysis of regulatory networks has *not* detected significant correlation between the robustnesses of multiple functions (Martin and Wagner [2008]). Most probably, whether robustness trade-offs are detected or not crucially depends on the observed features of the system and on the perturbation. In this thesis, a potential robustness trade-off is observed in the examined models concerning the features period and oscillation probability of the oscillators. In fact, for positive feedback or saturating kinetics in the chain models, increased period sensitivities as well as oscillation probability is obtained. This means that under these conditions, the periods are less robust while the occurrence of the oscillations is more robust. In contrast, a negative feedback-only model displays lower oscillation probability as well as lower period sensitivities than a positive feedback model (similarly found in Tsai et al. [2008]), and linear kinetics models can exhibit lower oscillation probabilities as well as lower period sensitivities than the according models with saturating kinetics (as in the negative feedback chain model in this work, section 6.4). Thus, the two features "oscillation probability" and "period robustness", both against parameter perturbations, certainly are not both high at the same time for any of the examined models, which implies the potential existence of a certain trade-off if both features are required to be robust.

Still, most of the biological oscillators and in particular circadian oscillations which require a stable period rely on positive-plus-negative feedback design which favors higher period sensitivities. This design seems most advantageous, and in fact, a positive feedback has been shown to increase the robustness of the period and the amplitude towards molecular noise (e.g. Gonze et al. [2002], Gonze and Hafner [2011]). Two possibilities to resolve this apparent contradiction are the following. First, all models examined here which deliver increased median period sensitivities also exhibit few, but at least some parameter sets with a low period sensitivity (sections 4.3, 5.3, 6.3.2, 7.2, Appendix section A.5). Consequently, one possibility to thus use a positive feedback structure and still obtain low period sensitivities may be that the values of the kinetic parameters are optimized towards low period sensitivity in nature. Therefore, the positive-plus-negative feedback design might still be an option even if stable periods are required. A second possibility to obtain stable oscillatory period and high oscillation probability could be the establishment of compensatory mechanisms which stabilize the period or lead to increased oscillation probabilities (and thus no strong positive feedback has to be used). For example, recently, a temperature-compensated synthetic oscillator relying on temperature sensitivity of a repressor protein has been engineered (Hussain et al. [2014]).

Additionally, coupling of oscillators can synchronize oscillations and increase the oscillation probability of the single oscillator (as observed in the SCN, Gonze et al. [2005], Locke et al. [2008]). Further compensatory mechanisms for steady state sensitivities comprise redundancy or degeneracy of structural motifs (Stelling et al. [2004b], Baggs et al. [2009], Shinar and Feinberg [2010], Steuer et al. [2011]), or modularity, i.e. encapsulation of structures leading to different features (Stelling et al. [2004b], Csete and Doyle [2002]). Similar mechanisms to compensate fluctuations in period or amplitude of oscillatory processes upon perturbation might be found in the future.

For the whole analysis presented in this work, ODE models are used. For circadian oscillations, this has been found to be appropriate, since robust oscillations have been detected for molecule numbers of $10 - 100$ (Gonze et al. [2002]). For calcium oscillations, ODE models are considered to represent the macroscopic level (Dupont et al. [2011]). Thus, the approach using ODE models can be considered largely appropriate in this thesis. Main parts of this thesis are dedicated to the analysis of chain models as prototype oscillator models. Neglecting features of the biological processes serves for keeping the models quite simple in order to be able to argue on few structural features. Already for the quite small calcium models examined in section 7.2, the complexity is high and makes in-depth analysis of the contribution of single structural features infeasible.

In the presented work, it is focused on perturbations of every single kinetic parameter in order to determine the period and amplitude sensitivity to avoid bias on the choice of the particular parameter and to keep the analysis as simple as possible. Investigating instead the effect of altering combinations of parameters or all parameters simultaneously would be appealing, as in the approaches proposed in Ihekwaba et al. [2005], Hafner et al. [2009], Dayarian et al. [2009], Zamora-Sillero et al. [2011]. Additionally, another possible perturbation is molecular noise, which requires the use of stochastic modeling. Positive feedback has been shown to reduce the sensitivity of the period and amplitude with respect to molecular noise (Gonze and Hafner [2011]) and it would be highly interesting to similarly examine the effects of matter flow between species and saturating kinetics on the robustness of the period and amplitude to molecular noise.

Other structural principles than the type of the feedback, matter flow or kinetics are also possible to address with the approach given in this work. As mentioned in section 7.2, the effects of the position of feedback loops and the strength of feedback loops (which characterizes for example the difference between short loops resembling conversion cycles consisting of reactions with linear opposed to saturating kinetics) on period and amplitude sensitivity are interesting prospective topics. Similarly, conservation conditions could possibly shape the sensitivity of oscillatory properties and deserve further investigations. Additionally, also time delay is an essential feature of oscillations (Wagner [2005], Novak and Tyson [2008], Hogenesch and Ueda [2011]). Increased time delay via ubiquitinations leads to increased period and amplitudes in different basic oscillator models (Xu and Qu [2012]). Its influence on the period sensitivities of the chain models has been examined for one parameter set by prolonging the chain (Wolf et al. [2005]). It has been found that an increase in chain length leads to a decrease in period sensitivity for both the negative and the positive feedback chain model (as also observed for increasing the chain length to five in this work, Appendix section A.2.3). Time delay could also be included explicitly requiring the use of delay differential equations for the analysis, or via switch-like responses or saturated degradation (Mengel et al. [2010]). This last aspect has been examined here (section 6.3.2), with the contrary outcome compared to prolonging the chain. The results indicate that the effect of time delay on the period and amplitude sensitivities may depend on how exactly it is realized. Consequently, a comparison between the effects of different implementations of increased or decreased time delay on the period and amplitude sensitivities could deliver interesting insights.

On the basis of the present study, predictions can be made concerning design principles to obtain, for example, oscillations with a sensitive or robust amplitude or period. Of course, not only forecasting predictions are important, but they also need to be tested. In the last years, synthetic biology has developed fast, and several synthetic oscillators with different feedback loops have been realized (e.g. Stricker et al. [2008], Tigges et al. [2009]). Recently, also feedback loops which can be regulated based on in silico methods have been established (reviewed in Chen et al. [2013]). For introducing saturating kinetics especially in degradation reactions, a tunable mechanism of overloading the degradation machinery by several proteins yielding the same degradation tag as instantiated in Cookson et al. [2011] might be possible. Thus, the research in synthetic biology already yields sufficient means to test some of the predictions as for example the role of saturation in the synthetic oscillators (section 7.3.2) derived here.

Along this line, according to the results of this thesis, it is strongly suggested to carefully consider especially the types of kinetics employed in biological models and the use of similar degradation tags for many proteins in synthetic biology, especially if the system is required to exhibit stable periods. The importance of saturation should be kept in mind also for experimental situations. If doing e.g. overexpression experiments, one should be aware that thereby the saturation of a reaction might be changed compared to the wild-type situation, and hence also oscillatory properties and steady states could display a much larger (or lower) sensitivity. Measurement techniques allowing for a better resolution of concentrations are hence desirable whenever the exact behavior of oscillatory and steady state properties needs to be determined.

A. Appendix

A.1. Robustness measure and methods

A.1.1. Influence of the parameter perturbation on the sensitivity

The sensitivity coefficients and overall sensitivity as used in this work (section 2.2, Equations 2.5, 2.6) are defined *a priori* for any parameter perturbation. For the sensitivity analysis in this work, the single parameter perturbation is chosen to be an increase by 2%. The effect of using different values for the parameter perturbation Δp or a different perturbation direction on the obtained sensitivity distribution is assessed here. As example models, a calcium oscillation model (Goldbeter et al. [1990], equations in section A.7.1) and a circadian oscillations model (Becker-Weimann et al. [2004], equations in section A.7.1) are used.

The increase of the parameters is varied ($\Delta p = 1\%$, $\Delta p = 5\%$, $\Delta p = 10\%$) and a sensitivity analysis is performed up to finding 250 parameter sets for which the sensitivity could be determined for each of the two models. Figure A.1 A shows an example comparing the sensitivity distributions for the two models for $\Delta p = 2\%$ (blue and red, white symbols for the median values) and $\Delta p = 10\%$ (black dots, black diamonds as median values). Not only those two, but all obtained sensitivity distributions yield similar characteristics. This is visualized in box-plots in Figure A.1 B for the period sensitivity and A.1 C for the amplitude sensitivity. All sensitivity distributions obtained from alterations in the parameter perturbation value are compared to those with $\Delta p = 2\%$ by a MWU test (Table A.1). Except for the period sensitivity distribution with $\Delta p = 10\%$ for which a significant decrease is stated (MWU p-value of $4.4 \cdot 10^{-9}$, median changes from 0.137 to 0.130), all other amplitude sensitivity and period sensitivity distributions do not differ significantly from those with $\Delta p = 2\%$. However, the number of parameter sets needed to be sampled is also decisively increased for $\Delta p = 10\%$ compared to the smaller perturbation sizes (from around 6900-7800 to 9300 or from 55000-68500 to 74500, see the last column of Table A.1) which slows the calculation process of the sensitivity analysis. Therefore, only small changes in the parameters are considered as perturbation. Then, it can be concluded that the particular choice of the parameter perturbation seems not to bias general results of the sensitivity analysis.

In order to avoid extended simulation effort in the sensitivity analysis, after sampling a parameter set, each single parameter is perturbed only in one direction with

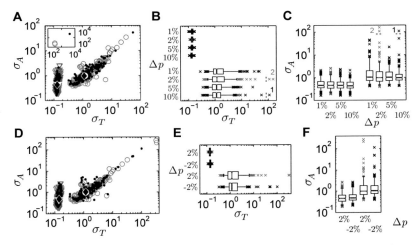

Figure A.1.: Impact of the parameter perturbation value on the sensitivity results for a calcium oscillations model (Goldbeter et al. [1990]) and a circadian oscillations model (Becker-Weimann et al. [2004]). A: Results of the sensitivity analysis for both models with $\Delta p = 2\%$ (blue: calcium model, red: circadian model, white symbols: median values) or $\Delta p = 10\%$ (black dots, black diamond: median values). B and C: Period and amplitude sensitivity distributions, respectively, as box-plots for different parameter perturbation values ($\Delta p = 1\%, 2\%, 5\%, 10\%$). The according distribution for $\Delta p = 2\%$ is depicted in color (red: circadian model, blue: calcium model). D: Results of the sensitivity analysis for both models with $\Delta p = 2\%$ (red: circadian model, blue: calcium model, white symbols: median values) or $\Delta p = -2\%$ (black dots, black diamond: median values). E and F: Period and amplitude sensitivity distributions, respectively, from panel D as box-plots.

Table A.1.: Results for the MWU test comparing the sensitivity distributions of the calcium oscillations model and the circadian oscillations model given in Figure A.1 for different parameter perturbations Δp. All sensitivity distributions are compared to the sensitivity distribution with $\Delta p = 2\%$. Given are the medians (μ), the MWU p-values (MWU p), U, z-values (z-val) and the total number of parameter sets sampled in each sensitivity analysis to obtain 250 sensitivity data points (# sets). Every distribution is composed of 250 data points, i.e. $n_X = n_Y = 250$. Data corresponds to Figure A.1.

Δp	period sensitivity				amplitude sensitivity				# sets
	μ	MWU p	U	z-val	μ	MWU p	U	z-val	
calcium oscillations model (blue)									
$\Delta p = 1\%$	1.161	0.29	-0.56	30347	1.034	0.035	-1.81	28325	7365
$\Delta p = 2\%$	1.183				0.984				6781
$\Delta p = 5\%$	1.146	0.15	-1.05	29550	1.015	0.19	-0.88	29836	7764
$\Delta p = 10\%$	1.133	0.19	-0.88	29830	1.034	0.38	-0.29	30777	9341
$\Delta p = -2\%$	1.179	0.49	-0.029	31203	1.026	0.11	-1.25	29225	7671
circadian oscillations model (red)									
$\Delta p = 1\%$	0.138	0.12	-1.16	29369	0.455	0.18	-0.93	29745	55032
$\Delta p = 2\%$	0.137				0.454				66388
$\Delta p = 5\%$	0.136	0.053	-1.62	28634	0.451	0.46	-0.11	31071	68491
$\Delta p = 10\%$	0.130	$4.4 \cdot 10^{-9}$	-5.75	21960	0.426	0.268	-0.64	30222	74777
$\Delta p = -2\%$	0.141	0.003	-2.75	26814	0.476	0.039	-1.76	28403	65630

a fixed percentage change of $+2\%$. Here, it is shown that if choosing the opposite perturbation direction and instead decreasing the parameter values by 2%, the obtained sensitivity distributions remain similar (Figure A.1 D-F). In the example models of Goldbeter et al. [1990] and the circadian oscillation model of Becker-Weimann et al. [2004], it is sampled only up to a number of 250 parameter sets for which the sensitivity could be determined for the parameter perturbation being $\Delta p = 2\%$ (blue and red distributions, white symbol as median) or $\Delta p = -2\%$ (black dots, black diamond as median). The obtained sensitivity distributions have similar characteristics (Figure A.1 E for the period sensitivity, F for the amplitude sensitivity). The median period sensitivities are very similar changing only by 0.034% from 1.183 to 1.179 for the calcium model and by 2.9% from 0.137 to 0.141 for the circadian model. The p-values of the MWU tests are 0.49 indicating highly insignificant differences for the calcium model and 0.003 for the circadian model indicating that the alteration is significant (Table A.1). For the amplitude sensitivities, median values change by 4.3% from 0.984 to 1.026 for the calcium model and by 4.8% from 0.454 to 0.476 for the circadian model. The p-values of the MWU tests are both slightly above 0.03 indicating no significant differences for the amplitude sensitivity distributions. Hence, whether the parameter values are perturbed by increasing or decreasing them seems to only affect slightly the results of the sensitivity analysis for this random sampling approach. It yields no substantial gain in information to justify the extended computational effort needed for checking for changes in both directions.

A.1.2. Influence of the definition of the sensitivity measure on the sensitivity results

In the approach of this work, the sensitivity is calculated as quadratic mean over the sensitivity coefficients of the nl-parameters and the rate coefficients of the models. Other measures of sensitivity based on sensitivity coefficients are of course possible. With the help of two example models, a circadian oscillation model by Becker-Weimann et al. [2004] and a calcium oscillation model by Goldbeter et al. [1990], the effects of two different definitions of the overall sensitivity from the sensitivity coefficients of the parameters on the sensitivity distributions are assessed (Figure A.2). The data for the single sensitivity coefficients from the original sensitivity analysis of the models are used for the calculation.

In Figure A.2 A, the results for the original sensitivity measure (circadian model in red, calcium model in blue, white symbols for the median values) are compared to results when calculating the quadratic mean only over the sensitivity coefficients of the rate coefficients and neglecting the nl-parameters (black dots, black diamond for the median values). For the calcium model, the median sensitivity values are decreased by 18% and 40% for period and amplitude, respectively. The median period sensitivity of the circadian model is increased by 11%, the amplitude slightly, but significantly decreased by 1.1% (p-value of the MWU test is 0.0056). These tendencies can also be observed in the according box-plots on top and to the right of the panel.

In Figure A.2 B, the original results from the sensitivity analysis (circadian model in red, calcium model in blue, white symbols for the median values) are compared

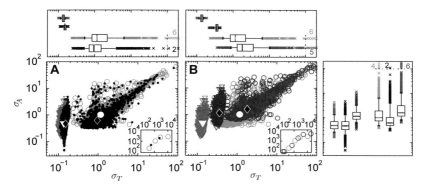

Figure A.2.: Impact of the definition of the sensitivity measure on sensitivity results for a circadian oscillation model (citebecker-weimann2004a) and a calcium oscillation model (Goldbeter et al. [1990]). The original sensitivity distributions of the circadian model are depicted in red triangles, those of the calcium model in blue circles. Median values are coded in white symbols. A: Sensitivities only summing over the sensitivity coefficients of the rate coefficients are given by black dots, black diamond for median values. B: Sensitivities summing over the three largest absolute sensitivity coefficients of all parameters are given by dark red triangles (circadian model) or dark blue circles (calcium model). Black diamonds are used for median values.

to the sensitivity values when summing over the largest three sensitivity coefficients (dark red for the circadian model, dark blue for the calcium model, black diamond for the median values). For both models, there is a visible increase for the period sensitivity as well as amplitude sensitivity values. The median sensitivities are increasing to around 2.5-fold for the circadian model and only to 1.5-fold for the calcium model. Thereby, the distance between the sensitivity distributions of the two models gets smaller. Still, the overall tendencies in comparing the sensitivities of the models remain approximately similar (section 3.2.2).

A.1.3. Influence of the sampling interval on the sensitivity results

In the parameter sampling process of the sensitivity analysis in this work, the steady state concentrations, steady state flows and nl-parameters are sampled in the interval $(10^{-3}, 10^3)$ (section 2.3.1). The effect of choosing a different interval on the results of the sensitivity analysis is shown here with an example model describing calcium oscillations by Goldbeter et al. [1990]. The resulting sensitivity distributions obtained using the interval $(10^{-3}, 10^3)$ are given in Figure A.3 (blue circles, large white circle as median value) together with the results for sampling in the interval $(10^{-1}, 10^5)$ instead (black dots, diamond for median values).

In panel A, period T versus period sensitivity σ_T are shown. The median period is decreased from 1.71 to 1.31 for using the altered sampling interval. The p-value of the MWU test is 0.033 indicating that the period distributions for the two different sampling intervals do not differ significantly.

In panel B, amplitude A vs. amplitude sensitivity σ_A is depicted. For the amplitude A, the median value is even increased by two orders of magnitude from 2.26 to 176 (Figure A.3 C), the change is significant (p-value of the MWU test is zero). This is as expected since the steady state concentrations are directly sampled from the sampling interval enabling larger amplitudes for the system if the sampling interval is shifted to higher values.

In panel C, amplitude sensitivities σ_A vs. period sensitivities σ_T are depicted. The

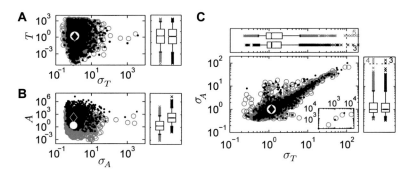

Figure A.3.: How the choice of the sampling interval influences the results of the sensitivity analysis for a calcium oscillation model (Goldbeter et al. [1990]). Blue circles, white circle as median: original sampling procedure with the interval $(10^{-3}, 10^3)$, black dots, black diamond as median: results when sampling in $(10^{-1}, 10^5)$. A: Period length T versus period sensitivity σ_T. B: amplitude A versus amplitude sensitivity σ_A. C: Period sensitivity σ_T versus amplitude sensitivity σ_A.

distributions for the different sampling intervals appear to be very similar (box-plots in panel C, detailed characteristics in Table A.2). Indeed, the p-values of the MWU tests are 0.09 and 0.17 for period sensitivity and amplitude sensitivity, respectively, for comparing the different sampling intervals (Table A.3). Hence, although the obtained amplitude values are decisively changed, neither the period sensitivity distributions nor the amplitude sensitivity distributions display significant differences if altering the sampling interval. Hence, the sensitivity analysis can be considered quasi independent from the particular choice of the borders of the sampling interval.

Table A.2.: Sensitivity values of a calcium model (Goldbeter et al. [1990]) for different sampling intervals: the standard interval $(10^{-3}, 10^3)$ or the altered interval $(10^{-1}, 10^5)$.

id	sensitivity	med	95% CI	$[Q_{25},Q_{75}]$	$[Q_5,Q_{95}]$	90%R	R_S	no.
$(10^{-3},10^3)$	period	1.158	[1.14,1.17]	[0.85,2.22]	[0.53,6.45]	5.92	0.82	$7.1 \cdot 10^5$
	ampl.	1.014	[1.00,1.03]	[0.74,1.94]	[0.46,6.54]	6.08		
$(10^{-1},10^5)$	period	1.166	[1.15,1.19]	[0.89,2.09]	[0.55,6.30]	5.75	0.83	$6.7 \cdot 10^5$
	ampl.	1.018	[1.01,1.03]	[0.77,1.87]	[0.47,6.48]	6.01		

Table A.3.: Results for the MWU test comparing the sensitivity distributions of a calcium oscillations model (Goldbeter et al. [1990]) for different sampling intervals: The standard interval $(10^{-3}, 10^3)$ or the interval $(10^{-1}, 10^5)$.

id_X	id_Y	sensitivity	MWU p	n_X	n_Y	U	z-val
$(10^{-3},10^3)$	$\Leftrightarrow$ $(10^{-1},10^5)$	period	0.09	2500	2500	3056538	-1.3414
		amplitude	0.17	2500	2500	3077044	-0.9396

A.1.4. Values obtained for special rate equations by the sampling method

During the sampling approach, the nl-parameters and steady state concentration values are sampled in the same interval following a log10-uniform random distribution. Here it is shown which steady state values result for Michaelis-Menten reaction terms. Therefore, a single Michaelis-Menten term with constant maximal reaction velocity $V = 1$ given by $\nu^0 = \frac{S^0}{S^0+K_M}$ is sampled. Sampled are 100000 sets with the sampling method described in 2.3.1. The results are summed up in Figure A.4.

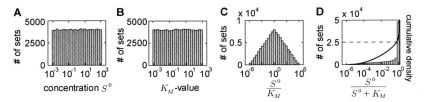

Figure A.4.: Values obtained by the sampling method used for 100000 sampled parameter sets. A: Histogram for a sampled steady state species concentration S^0. B: Histogram for a sampled K_M-value. C: Calculated value of the ratio between steady state concentration S^0 and K_M-value. D: Calculated Michaelis-Menten term $\frac{S^0}{S^0+K_M}$ together with the estimated cumulative probability function. A cumulative probability of 0.5 is indicated by the dashed line.

In panels A and B, histograms for the steady state substrate concentration S^0 and the K_M-value of the reaction are shown. It can be seen that both distributions follow a log10-uniform distribution in the sampling interval $(10^{-3}, 10^3)$. Panel C shows that most often, steady state substrate concentration S^0 and K_M-value are in the same order of magnitude. Their ratio takes values between 10^{-6} and 10^6 and is similarly often below and above 1.

The results of the sampling for the distribution of the Michaelis-Menten term is given as a histogram and cumulative density in panel D. From the histogram it is concluded that the Michaelis-Menten term takes values close to 1 (near saturation) most often. Nevertheless, the cumulative density data reveals that half of the samples take values below 0.5 (intersection between the dashed line and the cumulative density plot), the other half above 0.5 clarifying that non-saturated conditions are not at all underrepresented during the sampling procedure. Therefore, both saturated and non-saturated conditions are met sufficiently often for considering the used sampling procedure being fair to this regard.

A.1.5. Median vs. arithmetic mean

Instead of the arithmetic mean, this work employs the median value. The reason for this is illustrated with the help of an example of 100 data points generated in the course of a sensitivity analysis in this work (Figure A.5). The data points are indicated by crosses. They cluster at values between 0.5 and 1 exhibiting few extreme values. In Figure A.5 A, it is visible that the mean value, which is given by the circle, is strongly influenced by these extreme values and no longer reflects the actual properties of the data sample. It does not at all display the central tendency of the data, but it is only near three of the data points. The median is given by the square in Figure A.5 A, and it reveals to better show the central tendency of the data.

Additionally, in contrast to the arithmetic mean and associated measures, the median and the 90%R have the advantage of being robust with respect to extreme values. This means that if changing extreme values (to values more or less extreme) or if leaving them out, the median value and 90%R change only slightly whereas the mean and standard deviation would be considerably different. For the median and mean, this is also illustrated in Figure A.5. In panels B and C, the different behavior of the mean and the median values are shown if changing only one extreme value (B) or if leaving it out (C). By both procedures, the median is only slightly affected. It resides at a value of 0.555 for A and B and at a value of 0.554 for C. In contrast, the mean is significantly altered (A: 5.2, B: 1.8, C: 0.77).

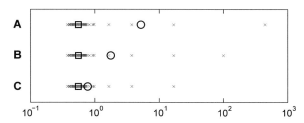

Figure A.5.: Comparison of mean versus median value. In A, a data sample with 100 data points is displayed by crosses. The circle shows the arithmetic mean value, the square the median value. In B, the data point with the highest value is changed from 440 to 100. In C, it is left out for determination of mean and median values.

A.2. The effect of feedbacks on sensitivity

A.2.1. Definition of the feedback strength and stability of the steady state

There are different definitions of feedback strength employed in the literature. The definition of the feedback strength as used in this work for the negative feedback chain model is $\xi = \frac{\partial log(\nu_2)}{\partial log(S_4)}$ (Equation 4.4). This is directly related to the percentage of unstable steady states obtained in the sampling procedure and thus also to the number of sets yielding sustained oscillations as can be seen in Figure A.6 A (color code: percentage of unstable steady states, white level lines: feedback strength ξ). In contrast, other measures of the feedback strength do not show this relation. Neither for the feedback strength being the feedback term at the examined steady state, i.e. $\overline{\xi} = \frac{1}{1+(S_4/kn_1)^n}$ (as e.g. in Xu and Qu [2012]), nor for the feedback strength being $\hat{\xi} = 1/kn_1$ (as e.g. in Qu and Vondriska [2009], Nguyen [2012]) the feedback

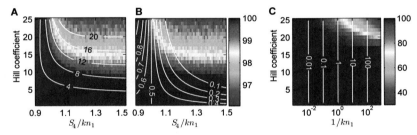

Figure A.6.: Comparison of different definitions of feedback strengths and stability for the negative feedback chain model. In all three panels, the percentage of stable steady states obtained is indicated in color-code. The level lines encode the feedback strength according to different definitions, A: the elasticity of the feedback reaction $\xi = \nu_2 \frac{\partial log(\nu_2)}{\partial log(S_4)}$ (the definition used in this work), B: the value of the feedback term at the steady state $\overline{\xi} = \frac{1}{1+(S_4/kn_1)^n}$, C: the inverse of the inhibition constant $\hat{\xi} = 1/kn_1$. Stability percentages are obtained as follows. A, B: The interval $[0.9, 1.5]$ for S_4/kn_1 is subdivided in intervals of length 0.01. In each of the intervals, 10000 parameter sets are sampled and the stability of the steady state is determined (compare section 2.3.2). C: 1000 parameter sets are sampled for $1/kn_1$ in each interval $10^{[a.a+0.05]}$ for $a = -3, -2.95, \ldots, 2.95$, the stability is determined for each sampled parameter set.

strength directly indicates how many unstable steady states occur (Figure A.6 B, C). Thus, the feedback strength as defined in this work seems most appropriate for the chain models employed here.

A.2.2. Positive feedback term and Hill coefficient

For the choice of the feedback term in the chain model, it has been discussed that often for the positive feedback, rather the term

$$\overline{fb_p} = \frac{S_4^n}{S_4^n + kn_1^n},$$

which is the additive inverse to the negative feedback term, is chosen instead of the herein employed $fb_p = \frac{S_4^n + kn_1^n}{S_4^n}$ being the multiplicative inverse. The results for a sensitivity analysis of the chain model employing $\overline{fb_p}$ as positive feedback term, keeping $n = 2$, is shown in Figure A.7 B (black dots, black diamond as median values), the characteristics of the sensitivity distributions are summed up in Table A.4. Both period and amplitude sensitivities are strongly increased for the altered feedback implementation. The median period sensitivities and median amplitude sensitivities rise by a factor of 4 and 23 from 0.68 to 2.95 and from 0.57 to 12.97, respectively (Table A.4). Hence, the tendency to display higher period sensitivities for the positive feedback chain model compared to the negative feedback chain model is still captured using the other feedback term while the conclusion to obtain lower amplitude sensitivities is no longer valid. Together with the strong increase in the median values, also the absolute variability of both period sensitivity and amplitude sensitivity is increased. However, neither low period sensitivities nor low amplitude sensitivities can be obtained with the positive feedback chain model with altered feedback term. All in all, the overall characteristics of the sensitivity distributions are considerably different if using the feedback term $\overline{fb_p}$ being the additive inverse to the Hill-inhibition negative feedback term. For the three reasons explained in section 4.2.1 and since the oscillation probability is decisively reduced (by a factor of approx. 40, $5.7 \cdot 10^6$ parameter sets needed to be sampled instead of $1.5 \cdot 10^5$ for the original positive feedback chain model), the original multiplicative inverse feedback term is preferred for further investigations of the positive feedback chain model.

In the chain models, the Hill coefficients are chosen to be the lowest numbers possible which allows for the occurrence of unstable steady states and hence for the existence of sustained oscillations, they are $n = 2$ and $n = 9$ for the chain model with positive and negative feedback, respectively (compare section 4.2.2). Here, for the positive feedback chain model with Hill coefficient of $n = 9$, a sensitivity analysis is performed (Figure A.7 B, black dots and black diamond for the results, Table A.4 for the characteristics of the sensitivity distributions). The in such a way altered positive feedback chain model displays a higher median period sensitivity but a lower median amplitude sensitivity than the positive feedback chain model with a Hill coefficient of $n = 2$ (0.78 and 0.49 for $n = 9$ compared to 0.68 and 0.57 for $n = 2$ for period and amplitude, respectively, Table A.4). For the variability, the 90%R for the period sensitivity distribution is found to be larger (2.81 for $n = 9$ compared to 1.99 for $n = 2$, Table A.4) whereas for the amplitude sensitivity distribution, it is considerably smaller (1.73 for $n = 9$ compared to 4.48 for $n = 2$, Table A.4). This indicates that

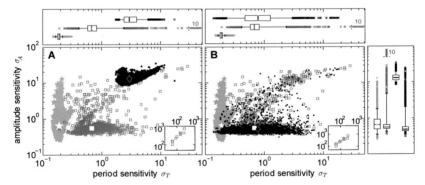

Figure A.7.: Impact of the positive feedback term and the Hill coefficient on the sensitivity of the chain model. Chain models with negative feedback (orange stars, white star as median), with positive feedback and Hill coefficient $n = 2$ (dark green squares, white square as median). Dark dots with black diamond as median values indicate the positive feedback chain model with feedback term $\overline{fb_p}$ and Hill coefficient $n = 2$ (A) and the positive feedback chain model with Hill coefficient $n = 9$ (B), respectively.

Table A.4.: Sensitivity distribution characteristics of the chain model with mass action kinetics, matter flow and positive feedback and a different positive feedback term ($\overline{\text{pos}}$) or a Hill coefficient of $n = 9$ (pos, $n = 9$) instead of $n = 2$. Data corresponds to Figure A.7.

id	sens.	μ	95% CI	$[Q_{25},Q_{75}]$	$[Q_5,Q_{95}]$	90%R	R_S	# sets
$\overline{\text{pos}}$	per.	2.95	[2.90, 3.00]	[2.41,3.94]	[1.99,6.25]	4.26	0.36	$5.7 \cdot 10^6$
	ampl.	12.97	[12.83,13.10]	[11.26,15.16]	[9.52,20.02]	11.50		
pos,	per.	0.78	[0.76,0.81]	[0.47,1.27]	[0.23,3.04]	2.81	0.09	$9.9 \cdot 10^5$
$n = 9$	ampl.	0.49	[0.48,0.49]	[0.44,0.57]	[0.40,2.13]	1.73		

for $n = 9$, the positive feedback chain model displays an even larger region than the model with $n = 2$ where the amplitude sensitivity remains fixed at low values while the period sensitivity is variable. The correlation between period sensitivity and amplitude sensitivity is close to zero indicating that both properties can be varied independently of each other by the choice of the parameter set. Therefore, the overall characteristics of the positive feedback chain model are also found for the herein examined different Hill coefficient. Additionally, although reaching down to lower values thereby causing a stronger overlap to the sensitivity values obtained for the negative feedback chain model, still the median period sensitivity of the positive feedback chain model with $n = 9$ is higher and the median amplitude sensitivity is lower than for the negative feedback chain model. Thus, the general differences observed when comparing the negative feedback chain model to the positive feedback chain model remain similar for the altered Hill coefficient.

A.2.3. Length of the feedback loop

The lengths of the feedback loops in the chain models can be altered for example by prolonging the chain. The sensitivities of the chain models of length five (see ODE system in section A.7.5) are given in Figure A.8.

The results of the sensitivity analysis of the negative feedback chain model of length five is given and compared to the results for the chain models of length four in Figure A.8 A. Although the period sensitivity distribution is wider than for the chain model with negative feedback of length four and therefore overlaps more with the period sensitivities obtained for the positive feedback chain model of length four, the median period sensitivity as well as median amplitude sensitivity are quite similar to those of the negative feedback chain model of length four.

Analogously, the sensitivities occurring for the chain model of length five with positive feedback are examined (Figure A.8 B, ODE system in section A.7.5)). The period sensitivity as well as amplitude sensitivity distributions resemble those found for the positive feedback chain model of length four. These results indicate that the length of the feedback loop might not be of capital importance for the period and amplitude sensitivities.

Table A.5.: Sensitivity distribution characteristics of the chain model of length five with mass action kinetics, matter flow and negative feedback (id "neg 5") or positive feedback (id "pos 5"). Data corresponds to Figure A.8.

id	sens.	μ	95% CI	$[Q_{25},Q_{75}]$	$[Q_5,Q_{95}]$	90%R	R_S	# sets
neg 5	per.	0.17	[0.17,0.18]	[0.16,0.28]	[0.14,0.48]	0.34	0.45	$2.0 \cdot 10^6$
	ampl.	0.71	[0.69,0.74]	[0.47,1.27]	[0.32,4.14]	3.82		
pos 5	per.	0.62	[0.61, 0.62]	[0.52,0.75]	[0.37,2.20]	1.83	0.34	$2.4 \cdot 10^5$
	ampl.	0.53	[0.52,0.53]	[0.47,0.61]	[0.39,5.94]	5.55		

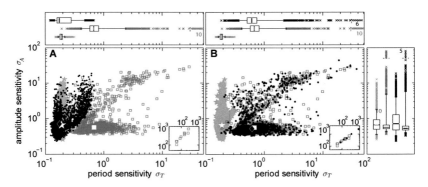

Figure A.8.: Impact of the length of the feedback loops on the sensitivity of the chain models. Chain models of length four: negative feedback (orange stars, white star as median), positive feedback (dark green squares, white square as median). Chain models of length five with negative feedback (panel A) or positive feedback (panel B): black dots, black diamond as median values. The accordingly colored box-plots capture the sensitivity distribution properties for the period (top) and amplitude (right).

A.3. The effect of matter flow on sensitivity

A.3.1. Transformation of models with different matter flow properties

A sensitivity analysis of all eight models with different matter flow properties as for negative feedback and all four oscillating models with positive feedback (Figure 5.2, section 5.2.1) is very extensive. As described in section 5.1, alterations of the matter flow properties between species are easily obtained and only minor changes have to be applied to the ODE system. Therefore, it might be possible to simplify the analysis. The analysis of one of the models might not be necessary if all its possible characteristics can also be found for another of the models. The sensitivity analyses of both models would hence yield similar results. In order to investigate whether this may happen, the relation of *transformation* of models is defined: model M1 can be transformed to model M2 if for each biologically feasible parameter set of model M1, there can be found a biologically feasible parameter set for model M2 such that the resulting ODE systems are the same. A parameter set is biologically feasible if it consists of positive parameter values only. ODE systems are the same if and only if they are yielding the same solutions. Therefore, the possible options for the infinite set of solutions of two models can be visualized as done in Figure A.9.

In order to practically approach the issue of transformation of models, two general facts deriving from properties of the derivative are needed: (i) If two ODE systems have the same Jacobian matrix and the right-hand side of the defining equations of the ODE system agrees at one point, the ODE systems are indeed the same and hence yield the same solution for any initial condition. (ii) Reversely, if two ODE systems are the same, their Jacobian matrices are the same and of course the right-hand sides of the ODE systems agree in one point. In summary, two ODE systems are the same and hence yield exactly the same solution if and only if (a) the right-hand side of the defining equations of the ODE system agrees in one point, and (b) they have the same Jacobian matrix.

For the chain models with different matter flow properties between species, the condition of agreement of the defining equations of the ODE at one point (condition (a)) is easily established if having the same rate coefficient k_1 for the production reaction ν_1. Then, for $S_i = 0$, $i = 1, \ldots, 4$, all reaction rates except for ν_1 are zero. Hence, the right-hand side of the defining equations is the vector $(k_1 \ 0 \ 0 \ 0)^T$ irrespective of the matter flow properties. To check whether the models can be transformed, the Jacobian matrices of the chain models with mass action kinetics remain to be compared (condition (b)). Figure A.10 A and B deliver the Jacobian matrices for the chain models C246 and C2--, respectively. The four entries in which the Jacobian matrices differ are j_{11}, j_{14}, j_{22} and j_{33}; they are marked by boxes. j_{11} and j_{14} are modulated only by the matter flow properties between species of ν_2, meaning that with ν_2 being a conversion, the terms $-k_2 \cdot fb$ and $-k_2 \cdot fb' \cdot S_1$ occur, respectively. j_{22} is modulated only by the matter flow properties of ν_4, meaning that for ν_4 being a conversion the term $-k_4$ occurs. j_{33} is modulated only by the matter flow properties of ν_6, meaning that with ν_6 being a conversion the term $-k_6$ occurs. Each change from conversion to regulated production or back in one of the reactions ν_2, ν_4 and ν_6 can hence be associated to the change of the respective entry or entries in the Jacobian matrix.

Comparing the Jacobian matrices, one can directly conclude that for parameter sets yielding oscillatory behavior, the ODE systems of two models differing in the

Figure A.9.: Visualization of transformation of models as used in this work. The red set defines the solutions to the ODE system of model M1 for any biologically feasible parameter set, the blue set the solutions to the ODE system of model M2. A: Neither model can be transformed to the other. There is even no overlap in the sets of solutions, meaning that no solution of the one model can be reproduced by the other. B: Neither model can be transformed to the other. There is an overlap in the space of possible solutions (violet area), but there are solutions in each model which are only occurring for their respective model. C: M2 can be transformed to M1. All possible solutions for model M2 also occur for model M1 for biologically feasible parameter sets, the solution space of M2 is a subset of the solution space of M1. M1 cannot be transformed to M2 since solutions of M1 exist which are not in the solution space of M2. D: M1 can be transformed to M2 and M2 can be transformed to M1. The solution spaces for biologically feasible parameter sets are identical.

A

$$J = \begin{pmatrix} \boxed{-k_3 - k_2 \cdot fb} & 0 & 0 & \boxed{-k_2 \cdot fb' \cdot S_1} \\ k_2 \cdot fb & \boxed{-k_4 - k_5} & 0 & k_2 \cdot fb' \cdot S_1 \\ 0 & k_4 & \boxed{-k_6 - k_7} & 0 \\ 0 & 0 & k_6 & -k_8 \end{pmatrix}$$

B

$$J = \begin{pmatrix} \boxed{-k_3} & 0 & 0 & \boxed{0} \\ k_2 \cdot fb & \boxed{-k_5} & 0 & k_2 \cdot fb' \cdot S_1 \\ 0 & k_4 & \boxed{-k_7} & 0 \\ 0 & 0 & k_6 & -k_8 \end{pmatrix}$$

Figure A.10.: A: Jacobian matrix for the chain model with mass action kinetics and matter flow between species for ν_2, ν_4 and ν_6 (model C246), B: Jacobian matrix for the chain model with ν_2, ν_4 and ν_6 being regulated productions and thus lacking matter flow between species (model C---). fb denotes the feedback term, fb' its derivative with respect to S_4. Entries differing between the two Jacobian matrices are marked by boxes.

matter flow properties of ν_2 can never be the same for any combination of biologically feasible parameter sets. This is due to entry j_{14} of the Jacobian matrix which changes from a non-zero value for ν_2 being a conversion (Figure A.10 A) to zero for lack of matter flow between species in this reaction (Figure A.10 B). The derivative of the feedback term is $fb' = \pm \frac{n_1 kn_1^{n_1} S_4^{n_1}}{(kn_1^{n_1} + S_4^{n_1})^2}$ where the sign depends on the feedback type. So for non-zero k_2, kn_1 and n_1, the possibility to obtain $j_{14} = 0$ would be to have either $S_1 \equiv 0$ or $S_4 \equiv 0$. The first option leads to $\nu_2 = 0$, the second keeps the feedback term constant and hence both cancel the feedback. Consequently, the negative feedback loop is lost and no oscillations can occur. Therefore, for any oscillatory solution, chain models with different matter flow properties between species of ν_2 do not yield the same ODE systems for biologically feasible parameter sets. Their solution spaces for biologically feasible parameter sets would be as depicted in Figure A.9 A or B. They cannot be transformed to each other

The case is different for the matter flow properties of reactions ν_4 and ν_6. Assume the ODE system for the model with matter flow in ν_4 or in ν_6 at a certain parameter set. Then, the ODE system for the model without matter flow in the respective reaction yields the same ODE system for the parameter set where only the degradation rate k_5 is substituted to $\hat{k}_5 := k_5 + k_4$ or $\hat{k}_7 := k_6 + k_7$, respectively, and the other parameters, especially also k_1 in order to fulfill condition (a), are kept the same. By this substitution, entry j_{22} or j_{33} of the Jacobian matrix of the model with matter flow gets $\hat{j}_{22}$ or $\hat{j}_{33}$, respectively, which equals the respective entry of the Jacobian matrix for the ODE system without matter flow. Hence, the Jacobian matrices of the models with and without matter flow are the same. If also $k_1 = \hat{k}_1$ remains the same, the ODE systems also agree at least in one point, and the ODE systems as well as the solution are the same. Thus, for every biologically feasible parameter set, a model with ν_4 or ν_6 (or both of them) being conversions can be transformed to a model with those reactions being regulated productions if the matter flow properties in all other reactions are the same. This may result in option C of Figure A.9 where the model lacking matter flow takes the role of model M2, since its solution space is a subspace of the solution space of the model with matter flow. It remains to be clarified whether also visualization D in Figure A.9 is appropriate for these models. This can be done with an example: Consider the model lacking matter flow only in ν_4. In order to convert this model to the model including matter flow in ν_4, only entry j_{22} of the Jacobian matrix would have to be altered keeping all other entries the same. Since the original value of j_{22} is $-k_5$ in the model lacking matter flow between species in ν_4 and $-\hat{k}_5 - \hat{k}_4$ for the model with ν_4 being a conversion, the parameters of the latter model (given with hats) would have to be adapted to satisfy $-\hat{k}_4 - \hat{k}_5 = -k_5$ or $\hat{k}_5 = k_5 - \hat{k}_4$. Since parameter $\hat{k}_4$ is fixed to $\hat{k}_4 = k_4$ by comparison of the entries j_{32} and $\hat{j}_{32}$ of the Jacobian matrices, it is $\hat{k}_5 = k_5 - k_4$. It is biologically reasonable to assume that the degradation rate coefficient k_5 of S_2 can be smaller than the reaction rate coefficient k_4 determining the reaction from S_2 to S_3. Whenever this happens, the required parameter set for the model with ν_4 being a conversion to obtain the same ODE system as the model with ν_4 being a regulated production yields a negative rate coefficient, $\hat{k}_5 = k_5 - k_4 < 0$. This does not constitute a biologically feasible parameter set and conversion of the models is not possible. Therefore, the case of differing matter flow properties between species in ν_4 and/or ν_6 is visualized in Figure A.9 C where the chain model with ν_4 and/or ν_6 being conversions takes the role of M1, the same model with lacking matter flow between species in ν_4 and/or ν_6 takes the role of M2, and option D from Figure A.9 is not possible.

In Figure A.11, the results of the preceding analysis are summed up. All eight chain models for each feedback type with different matter flow properties are given. An arrow from model M1 to model M2 means that M1 can be transformed to M2 or: For every biologically feasible parameter set of model M1, a biologically feasible parameter set in M2 can be found such that the ODE systems and hence also the solutions are the same. It can be observed that the process of this transformation is transitive, i.e. if model M1 can be converted to model M2 and model M2 can be converted to model M3, also model M1 can be converted to model M3. This fact can also be derived from set logic if applying the visualization from Figure A.9 C. In contrast, the process of transformation is not symmetric for the eight chain models examined here: If model M1 can be transformed to model M2, model M2 can even never be transformed to model M1 here. This is reflected by having exactly one arrow pointing only in one direction between any two models in Figure A.11. The separation of the four models to the left and the four models to the right in Figure A.11 are caused by the fact that chain models differing in the matter flow property of reaction ν_2 yield no possibility of transformation.

Be aware that the conclusions concerning the possibility to obtain the same ODE

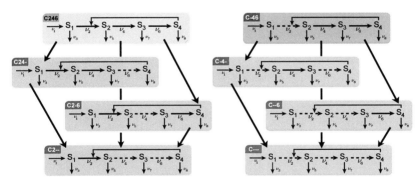

Figure A.11.: Transformation of chain models with mass action kinetics for different combinations of matter flow between species. For simplification, reactions regulated productions are indicated by dashed arrows. The arrow between models indicates that the model at the source can be transformed to the model at the head. This means that for every biologically feasible parameter set of the model at the source, it can be found a biologically feasible parameter set for the model at the head such that the two resulting ODE systems and in particular their solutions are identical.

systems for models with different matter flow properties are no longer valid if other kinetics than mass action kinetics are used for the reactions in the model.

A.3.2. Equivalent models may have different sensitivities

In section A.3.1, it has been shown that some of the chain models with mass action kinetics and different matter flow properties can be transformed into another chain model yielding the same ODE systems and solutions for different parameter sets. For the sensitivity analysis, it is of interest to determine whether two identical ODE systems deliver the same sensitivity of period and amplitude although the underlying models are different. In this case, it would be sufficient to examine the sensitivities of the chain model with only ν_2 being a conversion (model C2--) and that with all three reactions ν_2, ν_4 and ν_6 being a regulated production (model C---) since any other of the models can be transformed into them (see Figure A.11). One example pair of chain models is inspected in which the one can be transformed to the other and one particular parameter set each for which the ODE systems of the models are identical. It is checked whether they deliver the same period and amplitude sensitivities.

The starting point is chosen to be the chain model with negative feedback with all three reactions ν_2, ν_4 and ν_6 being conversions (model C246 in Figure A.11) at a particular parameter set found during the sensitivity analysis of the model (given in Table A.6, upper rows). The parameter set is changed in order to constitute the same ODE system for the chain model with negative feedback and only ν_2 being a conversion (model C2-- in Figure A.11) and is given in Table A.6 in the lower rows. The values that differ from those from the other model are high-lighted in gray. The steady state concentration values S_i^0, $i = 1, \ldots, 4$, of the species are the same for both models as well as the inhibition constant kn_1. The steady state flows ν_i^0 as well as the rate coefficients k_i differ in entry 5 and 7. This is caused by the need to include the decrease of species S_2 and S_3 by the conversions reactions ν_4 and ν_6 in model C246 into their degradation reactions $\hat{\nu}_5$ and $\hat{\nu}_7$ in model C2-- lacking matter flow in ν_4 and ν_6. It has to be $\hat{k}_5 = k_5 + k_4$, $\hat{k}_7 = k_7 + k_6$ if marking the parameters for model C2-- by the hat.

The oscillatory behavior of the models at the given parameter sets are depicted

Table A.6.: Parameters for the chain models with negative feedback and different matter flow properties yielding the same ODE system. Upper rows: Model C246 with ν_2, ν_4, ν_6 being conversions. Lower rows: Model C2-- including matter flow in ν_2 only. Given are the values for the steady state concentrations S_i^0, the nl-parameter kn_1, the steady state flows ν_i^0 and the rate coefficients k_i (which can be calculated from the other values). Values are rounded. Differences are high-lighted in gray.

model	S_1^0	S_2^0	S_3^0	S_4^0	kn_1		
C246	0.0284	2.81	0.683	0.664	0.0544		
C2--	0.0284	2.81	0.683	0.664	0.0544		

model	ν_1^0	ν_2^0	ν_3^0	ν_4^0	ν_5^0	ν_6^0	ν_7^0	ν_8^0
C246	77.8	0.687	77.1	0.167	0.520	0.1323	0.0344	0.132
C2--	77.8	0.687	77.1	0.167	0.687	0.1323	0.1667	0.132

model	k_1	k_2	k_3	k_4	k_5	k_6	k_7	k_8
C246	77.8	$1.47 \cdot 10^{11}$	$2.71 \cdot 10^3$	0.05937	0.18518	0.1937	0.0504	0.199
C2--	77.8	$1.47 \cdot 10^{11}$	$2.71 \cdot 10^3$	0.05937	0.24455	0.1937	0.2441	0.199

in Figure A.12 A for model C246 and B for model C2--. Both implementations lead to exactly the same oscillations. The absolute sensitivity coefficients for both parameter sets are given in panels C for period sensitivity and D for amplitude sensitivity. The values for the chain model C246 is shown in white stars, for the chain model C2-- in black circles. First of all, for all parameters not included in reactions ν_4 - ν_7 (rate coefficients 1, 2, 3, 8 and kn_1) the period sensitivity coefficients as well as the amplitude sensitivity coefficients are similar for both matter flow implementations. This is as expected since those reactions are not affected by the differing matter flow properties. For the period sensitivity coefficients (Figure A.12 C) of the rate coefficients 4 - 7 considerable differences exist. The chain model C246 (stars) exhibits significantly lower values for the parameters governing the degradation reactions (rate coefficients 5, 7) but higher values in the parameters for reactions ν_4 and ν_6 (rate coefficients 4, 6) than the model C2-- (circles). The reason for this lies in the increase of the values of the degradation rate coefficients for model C2-- caused by the degradation rate coefficients being the sum of the original degradation and conversion rate coefficients (see Table A.6). This results in larger absolute parameter changes for k_5 and k_7 for similar relative changes applied for the sensitivity estimation in chain model C2-- and hence enlarges also the period sensitivity coefficients. The unaltered conversion rate coefficients k_4 and k_6 are smaller in relation to their according degradation rates k_5 and k_7 in model C2-- than in model C246. Additionally, in model C2--, the rate coefficients k_4 and k_6 of the two reactions affect only the production of their product species but not the degradation of their source species. Hence, in model C2--, the period sensitivity coefficient for the two parameters are smaller compared to model C246. For the amplitude sensitivity coefficients which are given in Figure A.12 D, only the amplitude sensitivities of parameters k_6 and k_7 of model C2-- are found to be visibly different, in fact larger, than those of model C246.

All in all, model C2-- displays larger absolute period sensitivity coefficients than model C246. This could be explained by the fact that the decrease of species concentrations S_2 and S_3 in model C2-- is only regulated via the degradation rates

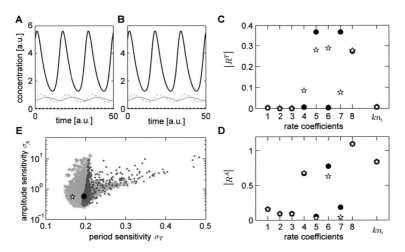

Figure A.12.: The effect of different matter flow implementations on sensitivity. A, B: Oscillations for the negative feedback chain model C246 (A) or C2-- (B) at the parameter set indicated in Table A.6. S_1: dashed black, S_2: black, S_3: dashed gray, S_4: gray. C, D: Absolute period (C) or amplitude (D) sensitivity coefficients for model C246 (white stars) and model C2-- (black circles). E: Overall sensitivities for the particular parameter set (model 246: white star, model 2: black circle) and the results of the sensitivity analysis with parameter sampling for the negative feedback chain model C246 (orange stars) or model C2-- (red dots).

(rate coefficients 5 and 7). Disturbing one of those parameters severely affects the velocity of the decrease of one of the species and hence disturbs the period of the cycle. In model C246, the species S_2 and S_3 are not only decreased by their degradation rate, but also by an alternative path. They are step by step processed to species S_4 and then degraded via reaction ν_8. Hence, the regulation of decrease of the species is not only partitioned between two but between four parameters, rate coefficients 4 - 7. This causes the overall period sensitivity which is the quadratic mean of the sensitivity coefficients of all parameters to be smaller for model C246. This can be observed in Figure A.12 E, where the overall sensitivity of the models at the parameters sets is depicted (white star: model C246, black circle: model C2--). Additionally, the results of the sensitivity analysis with parameter sampling is given for both models in Figure A.12 E. The period sensitivities of model C2-- do not reach as low as those obtained from model 246.

This elaboration shows that despite yielding possibly identical ODE systems, the chain models with different matter flow properties between species may exhibit different period and/or amplitude sensitivities. Therefore, it is not sufficient to examine the sensitivities of the two chain models into which all of the others can be transformed (see Figure A.11), but it is necessary to analyze every possible combination of matter flow properties and their effect on sensitivity.

A.3.3. Sensitivity analysis of the negative feedback chain models with different matter flow properties

In this section, the influence of the matter flow properties on the period and amplitude sensitivity of the chain model with negative feedback is examined. For all eight possible combinations of matter flow properties for reactions ν_2, ν_4, ν_6 (see Figure 5.2), a sensitivity analysis is performed for the model. The results are given by red squares and a white square for the median values in Figure A.13 and according values in section A.6 (Tables A.10, A.12). For comparison, in panels A-H in Figure A.13, the sensitivities obtained for the chain model with matter flow in all three reactions ν_2, ν_4, ν_6 (model 246) are given in orange stars and a black star for the median values. The box-plots of the period and amplitude sensitivity distributions are shown in panels I and J, respectively.

Period sensitivity

First, the effect of matter flow properties on the period sensitivity is described. All in all, it can be observed that for any alteration of matter flow properties, the median values of the period sensitivity change only slightly. Their values vary between 0.189 for model C246 and 0.200 for model C2-- (Table A.10). This constitutes a maximal

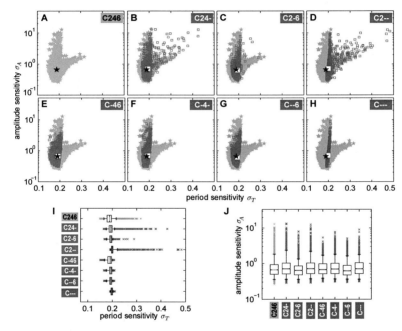

Figure A.13.: Impact of flow of matter on the sensitivity of the negative feedback chain model. A-H: Sensitivities for the chain model with matter flow for all three reactions ν_2, ν_4 and ν_6 are given as orange stars, black star as median values. Red squares (white square for median values) indicate the results for the chain model with negative feedback and matter flow between species in: ν_2, ν_4 (B); ν_2, ν_6 (C); ν_2 (D); ν_4, ν_6 (E), ν_4 (F), ν_6 (G). The results considering model C--- which lacks matter flow between species in all three reactions are shown in panel H. I and J: Box-plots of the period sensitivity and amplitude sensitivity distributions, respectively, for all models from A-H.

relative change of 5.8%. Despite the variation in the median values being slight and the strong overlap of the values of the period sensitivity distributions (Figure A.13 I), the differences in the period sensitivity distributions compared to the model C246 are significant as all p-values for the MWU tests comparing the period sensitivity distributions are below 0.0002 (see Table A.12).

The impacts of the matter flow properties on the period sensitivities differ for reaction ν_2 on the one side and reactions ν_4 and ν_6 on the other side. In Figure A.13 I, a shift in the median period sensitivity can be followed: The median period sensitivities increase with increasing number of reactions ν_4 and/or ν_6 being regulated productions instead of conversions (from model C246 to model C2-- and from model C-46 to model C---, respectively). In contrast to reactions ν_4 and ν_6, alterations in the matter flow property of reaction ν_2 can increase, decrease or even leave the median period sensitivity values constant (compare values for models C246 and C-46, models C2-6 and C--6 or models C24- and C-4-, respectively, in Table A.10), but all in all, they are only slightly affected by changing the matter flow property of ν_2 and vary by at most 1.1%.

The dispersion of the period sensitivity distribution as indicated by the 90%R also varies with altering the matter flow properties of the model, the highest value is 0.15 and occurs for model C246, the smallest is 0.05 for model C--- (Table A.10). Again, the matter flow properties of ν_4 and ν_6 have a similar, additive effect on the dispersion of the period sensitivity distribution; they increase the 5th percentiles (section 5.4.2, Q_5 values in Table A.10, lower tips of the whiskers in Figure A.13 I). The 90%R decreases with ν_4, ν_6 being regulated productions by at most 48% or 66% for models with ν_2 being a conversion (Figure A.13 A to D, upper four models in panel I) or for models with ν_2 being a regulated production (Figure A.13 E to H, lower four models in panel I), respectively. In contrast, lacking matter flow between species in ν_2 mainly reduces the Q_{95} (section 5.4.2), and up to 30% smaller 90%Rs are obtained (except for model C2-6 to model C--6, Table A.10).

Amplitude sensitivity

The matter flow properties also show to influence the amplitude sensitivities of the chain models with negative feedback. The median amplitude sensitivity values range from 0.62 to 0.71 for the eight examined models (Table A.10) which corresponds to a maximal variation of 14.5%. Again, the effect of altering the matter flow properties of single reactions can be assessed. All amplitude sensitivity distributions obtained for changes in matter flow properties of ν_4 and/or ν_6 are significantly different compared to that of model C246 (p-values are smaller than $1.3 \cdot 10^{-5}$, Table A.12). The models with ν_6 being a regulated production display increased median amplitude sensitivities (models C24-, C2--, C-4-, C---, values between 0.70 and 0.71, Table A.10) compared to the according chain models with ν_6 being a conversion (models C246, C2-6, C-46, C--6, values between 0.62 and 0.66, Table A.10), especially visible in Figure A.13 J. If lacking matter flow between species in ν_4 but not in ν_6 (compare models C246 and C2-6, models C-46 and C--6), the median amplitude sensitivities are reduced by 3.0% and 7.5%, respectively (Figure A.13 J, Table A.10). The matter flow property of ν_2 has no significant impact on the amplitude sensitivity distributions of the different negative feedback chain models (Figure A.13 A-D to E-H, all MWU p-values larger than 0.014, Table A.12).

The dispersion of the amplitude sensitivity distributions of the chain models with negative feedback measured by the 90%R varies between 1.0 and 2.02 (Table A.10). The dispersion increases most for only ν_6 being a regulated production (by 36%, model C246 to C24-) and decreases most for only in ν_4 being a regulated production (by 33%, model C246 to C2-6). Combining lack of matter flow between species in both ν_4 and ν_6, the 90%R still increases by 23% (model C246 to C2--).

In sum, the effects of replacing one of the conversions ν_2, ν_4 and ν_6 by regulated productions on the period and amplitude sensitivities of the negative feedback chain model are combined when lacking matter flow between species in multiple reactions. The median period sensitivity is only slightly changed due to alterations in matter flow properties while the median amplitude sensitivity is quite variable. Additionally, the dispersion of the period sensitivity decisively decreases the more reactions are regulated productions instead of conversions.

Sensitivity coefficients of the negative feedback chain models

In the main text, sections 5.4.1 and 5.4.2, the period sensitivity coefficients for the individual parameters of the negative feedback chain model are used to further investigate the origins of the increased period sensitivities and the reduced period sensitivity dispersion of the negative feedback chain model C--- compared to model

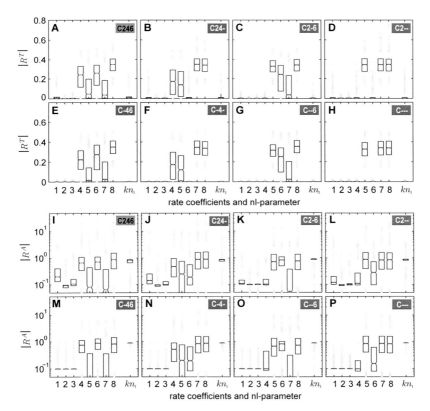

Figure A.14.: Sensitivity coefficients of the chain models with negative feedback and different matter flow properties of the reactions ν_2, ν_4 and ν_6. A-H and I-P: Box-plots of the distributions of the absolute period sensitivity coefficients $|R^T|$ and absolute amplitude sensitivity coefficients $|R^A|$, respectively, for each individual parameter of the models with matter flow between species in: ν_2, ν_4 and ν_6 (A, I); ν_2, ν_4 (B, J); ν_2, ν_6 (C, K); ν_2 (D, L); ν_4, ν_6 (E, M), ν_4 (F, N); ν_6 (G, O). Sensitivities of model C--- are given in panels H and P.

C246. Also the effects of the lacking matter flow between species for the single reactions of model C--- have been detected. Here, the absolute values of the period and amplitude sensitivity coefficients of all eight negative feedback chain models with different matter flow properties are given (Figure A.14).

Concerning the period sensitivity coefficients the results are in accordance with those elaborated in sections 5.4.1 and 5.4.2. If reaction ν_2 is a regulated production instead of a conversion, the upper limits of the $|R^T|$s of rate coefficients 1, 8 and of kn_1 are decreased (as the parameter sets with considerable higher period sensitivity than the median vanish), and the $|R^T|$s of rate coefficients $1 - 3$ are similar (Figure A.14 A-D compared to E-H). If reaction ν_4 is a regulated production instead of a conversion, the $|R^T|$s of rate coefficient 4 are decisively decreased, while those of rate coefficient 5 are strongly increased (Figure A.14 A to C, B to D, E to G, F to H). If reaction ν_6 is a regulated production instead of a conversion, the main effects on the $|R^T|$s are a decisive decrease for rate coefficient 6 and a decisive increase for rate coefficient 7 (Figure A.14 A to B, C to D, E to F, G to H). The effects of single disruptions of the matter flow are combined for multiple reactions lacking matter flow between species.

Concerning the amplitude sensitivities, the results are less homogeneous due to the matter flow properties of single reactions affecting the amplitude of single species rather than all species simultaneously, and thus the influences on the mean amplitude are blurred. However, the main effects of altering the matter flow properties of individual or combinations of reactions on the amplitude sensitivity coefficients are comparable to those for the period sensitivity coefficients (Figure A.14 I-P), especially for reactions ν_2 and ν_4. Lacking matter flow between species in reaction ν_2 has only slight effects on the sensitivity coefficients. It leads to smaller maximal $|R^A|$s in rate coefficients 1, 8 and in kn_1, and as for the period sensitivity, the $|R^A|$s of rate coefficients 1-3 are similar (Figure A.14 I-L compared to M-P). If reaction ν_4 is a regulated production instead of a conversion, as for the $|R^T|$s, the $|R^A|$s of rate coefficient 5 are decisively increased while those of rate coefficient 4 are decreased (Figure A.14 I to K, J to L, M to O, N to P). If reaction ν_6 lacks matter flow between species, as for the $|R^T|$s, the $|R^A|$s of rate coefficient 6 are decreased while those of rate coefficient 7 are increased (Figure A.14 I to J, K to L, M to N, O to P). Furthermore, the $|R^A|$s of rate coefficients 4 and 5 are affected. This is the reason why especially for altering the matter flow property of ν_6 the amplitude sensitivities are increased. Again, for multiple alterations in matter flow properties, the effects are combined (e.g. Figure A.14 I to P).

A.3.4. Sensitivity analysis of the positive feedback chain models with different matter flow properties

In this part, the sensitivities of the oscillatory characteristics of the models arising from manipulating the matter flow properties in the chain models with positive feedback are examined. If disrupting matter flow in ν_2, the positive feedback chain models do not yield oscillations (section 5.2.1). This is the reason why the sensitivity

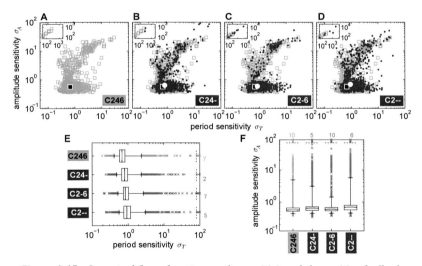

Figure A.15.: Impact of flow of matter on the sensitivity of the positive feedback chain model. A-D: Sensitivities for the chain model with matter flow for all three reactions ν_2, ν_4 and ν_6 are given as green squares, black square as median values. Dark green dots (white circles for median values) indicate the results for the chain model with positive feedback and matter flow between species in: ν_2, ν_4 (B); ν_2, ν_6 (C); ν_2 (D). E and F: Box-plots of the period sensitivity distributions and amplitude sensitivity distributions, respectively, for all models from A-D.

analysis in this section is restricted to the four models with ν_2 being a conversion and different matter flow properties in ν_4 and ν_6 (the four models to the left in Figure 5.2).

The results of the sensitivity analyses of the four models are shown in Figure A.15 and Tables A.11 and A.13. The analysis shows that if either reaction ν_4 or reaction ν_6 is a regulated production instead of a conversion, the median period sensitivity increases significantly, at least by 17.6% (Figure A.15 A to B or A to C, Table A.11, MWU p-values $< 10^{-64}$, Table A.13). If both reactions lack matter flow (Figure A.15 A to D), the rise in the median sensitivities is even more pronounced with an increase of 36%. The 90%Rs of the period sensitivity distributions of the chain models with positive feedback and different matter flow properties range from 1.89 for model C24- to 2.50 for model C2-- (Table A.11). This corresponds to a maximal relative change of dispersion of 32%.

A similar but slightly weaker effect of the matter flow properties of reactions ν_4 and ν_6 is observed on the median amplitude sensitivities. The median values lie in a range of 0.57 to 0.68 which corresponds to a maximal increase of 19% (from Figure A.15 A to D). Compared to model C246, all considered alterations in matter flow properties lead to a significant increase in the amplitude sensitivity distributions (Table A.13). The more reactions lack matter flow, the higher the median amplitude sensitivity becomes. The 90%Rs of the amplitude sensitivity distributions vary from 1.01 for model C2-6 to 5.20 for model C2-- (Figure A.15 C, D, Table A.11). For the change in dispersion, rather the 95th percentiles are responsible. The 5th percentiles are only slightly changed and range between 0.43 and 0.45 (Table A.11).

Sensitivity coefficients of the positive feedback chain models

In order to estimate the contribution of the individual parameters to the sensitivities and changes therein due to alterations in matter flow properties, the distributions of the absolute sensitivity coefficients of each model are analyzed and given as box-plots (Figure A.16).

In fact, the tendencies of alterations are quite similar for the period and amplitude sensitivity coefficients (compare Figure A.16 A-D to E-H). Since also the change in period and amplitude sensitivities upon alterations in matter flow properties of reactions ν_4 and ν_6 are similar for the positive feedback chain models, the induced changes are described in the following for absolute sensitivity coefficients $|R|$, meaning both $|R^T|$ as well as $|R^A|$. If ν_4 is a regulated production instead of a conversion, the $|R|$s of both rate coefficients 4 and 5 are decisively increased (Figure A.16 A to C, B to D, E to G, F to H). Similarly, for reaction ν_6 lacking matter flow between species, the $|R|$s of rate coefficients 6 and 7 are decisively increased (Figure A.16 A to B, C to D, E to F, G to H). Combining alterations in matter flow properties leads to additive combinations of their effects (Figure A.16 A to D, E to H). Indeed, the results confirm the lack of compensation for lacking matter flow between species detected in section 5.4.3 also for individual alterations of the matter flow properties of reaction ν_4 or ν_6.

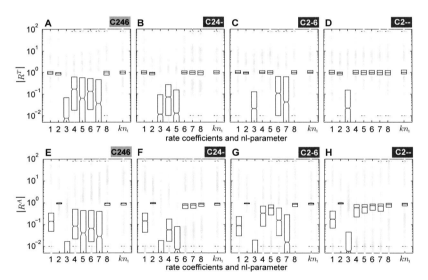

Figure A.16.: Sensitivity coefficients of the chain models with positive feedback and different matter flow properties of the reactions ν_4 and ν_6. A-D and E-H: Box-plots of the distributions of the absolute period sensitivity coefficients $|R^T|$ and absolute amplitude sensitivity coefficients $|R^A|$ for each parameter for the models with matter flow in: ν_2, ν_4 and ν_6 (A, E); ν_2, ν_4 (B, F); ν_2, ν_6 (C, G); ν_2 (D, H).

A.4. The effect of saturating kinetics on sensitivity

A.4.1. Derivation of Michaelis-Menten kinetics

Michaelis-Menten kinetics are originally derived as a simplification of an enzyme-driven reaction scheme (descriptions for example in Fell [1997], Cornish-Bowden [2004]). The reaction scheme using mass action kinetics is given in Figure A.17. Before being converted to a product, the substrate is reversibly bound to an enzyme. The formation of the enzyme-substrate complex ES, which is also called Michaelis complex, is performed by the reaction $\nu_1 = k_1 \cdot S \cdot E$ and the backward reaction by $\nu_{-1} = k_{-1} \cdot ES$. The reaction rate of conversion to the product depends on the concentration of enzyme-substrate complex: $\nu = k \cdot ES$. With the production reaction, the enzyme is released without experiencing any changes. The development in time of the enzyme-substrate complex is given by

$$\frac{dES}{dt} = \nu_1 - \nu_{-1} - \nu. \tag{A.1}$$

Including some assumptions, the system can be simplified as in Briggs and Haldane [1925]. Assuming constant total enzyme concentration $E_{tot} = E + ES$ and a quasi-steady state of the enzyme-substrate complex, $\frac{dES}{dt} = 0$, which is only possible if the substrate concentration is considerably larger than the enzyme concentration, the following can be derived:

$$0 = k_1 \cdot S \cdot E - k_{-1} \cdot ES - k \cdot ES$$
$$0 = k_1 \cdot S \cdot (E_{tot} - ES) - k_{-1} \cdot ES - k \cdot ES$$
$$k_1 \cdot S \cdot ES + k_{-1} \cdot ES + k \cdot ES = k_1 \cdot S \cdot E_{tot}$$
$$SE = \frac{k_1 \cdot E_{tot} \cdot S}{k_1 \cdot S + k_{-1} + k}.$$

Using $V_{max} = k \cdot E_{tot}$ and $K_M = \frac{k_{-1}+k}{k_1}$, one obtains the typical and known formula for a Michaelis-Menten reaction to the product:

$$\nu = k \cdot ES = k \cdot \frac{E_{tot} \cdot S}{S + \frac{k_{-1}+k}{k_1}} = V_{max} \frac{S}{S + K_M}. \tag{A.2}$$

Note that the rate coefficient of a reaction ν governed by Michaelis-Menten kinetics is most often denoted by V_{max} (also V_m or simply V as in this work). The reason is its function: Since the remaining part of the term $\frac{S}{S+K_M}$ is always smaller than 1, the rate coefficient determines the maximal possible velocity of the reaction ν. Note also that the maximal reaction velocity V_{max} depends linearly on the total enzyme concentration. This description is restraint to the mono-substrate cases. For multiple substrates, more possibilities occur (in dependence of the binding order or binding regime of the enzyme-substrate complexes, see e.g. Fell [1997], Cornish-Bowden [2004]).

$$S + E \xrightleftharpoons[v_1]{v_{-1}} ES \xrightarrow{v} P + E$$

Figure A.17.: Reaction scheme for the derivation of Michaelis-Menten kinetics. A substrate S binds reversibly to an enzyme E, the enzyme-substrate complex ES is converted to a product P.

A.4.2. Steady state response coefficients as in the main text

Derivation of R_ν for the Goldbeter-Koshland switch

The Goldbeter-Koshland switch is a system where a protein S_1 is reversibly converted to S_2 by enzyme-driven reactions (model scheme in Figure 6.4 A). Assuming Michaelis-Menten kinetics for both processes, the according ODE system is given by

$$\frac{dS_1}{dt} = V_2 \frac{S_2}{S_2 + K_2} - V_1 \frac{S_1}{S_1 + K_1}$$
$$\frac{dS_2}{dt} = V_1 \frac{S_1}{S_1 + K_1} - V_2 \frac{S_2}{S_2 + K_2}.$$

The parameters of reactions ν_1 and ν_2 are maximal reaction velocities V_1 and V_2 and the K_M-values K_1 and K_2 giving the binding affinities of the enzymes, respectively. Since $S_1 + S_2 = S_{tot} = const.$, for a given set of parameters and a defined total protein concentration S_{tot}, there is a unique positive steady state of the system which can be calculated analytically. Since $S_1^0 = S_{tot} - S_2^0$, and setting $\overline{K_i} = \frac{K_i}{S_{tot}}$, it is given by the equations

$$\frac{S_1^0}{S_{tot}} = \frac{\frac{V_1}{V_2}(1 + \overline{K_2}) + \overline{K_1} - 1 \pm \sqrt{(\overline{K_1} + 1)^2 + \frac{V_1}{V_2}(\overline{K_2} + 1)^2 - 2\frac{V_1}{V_2}(\overline{K_1} + \overline{K_2} - \overline{K_1}\overline{K_2} + 1)}}{2\frac{V_1}{V_2} - 2}$$

(A.3)

for $\frac{V_1}{V_2} \neq 1$, and

$$\frac{S_1^0}{S_{tot}} = \frac{\overline{K_1}}{\overline{K_1} + \overline{K_2}} \quad \text{for } \frac{V_1}{V_2} = 1.$$

In order to determine the response coefficients R_ν, the calculation of the steady state is not necessary. Indeed, already from the ODEs, it is

$$\frac{V_1}{V_2} = \frac{S_{tot} - S_1^0}{S_{tot} - S_1^0 + K_2} \frac{S_1^0 + K_1}{S_1^0}.$$

(A.4)

Hence, the reaction velocity ratio to obtain a percentage of $a \cdot 100$ percent unmodified protein at steady state (i.e. $S_1^0 = aS_{tot}$) is easily derived by inserting this value into Equation A.4:

$$\left(\frac{V_1}{V_2}\right)_a = \frac{S_{tot} - aS_{tot}}{S_{tot} - aS_{tot} + K_2} \frac{aS_{tot} + K_1}{aS_{tot}} = \frac{1 - a}{a} \frac{aS_{tot} + K_1}{S_{tot} - aS_{tot} + K_2}.$$

Using the definition of the general saturation index of reaction ν_i given by $sat_{\nu_i} = \frac{max(S_i^0)}{max(S_i^0) + K_i}$ and inserting that for each steady state transition, the maximal value $max(S_i^0)$ obtained (and also reached for $V_1/V_2 \to 0$) for S_1 is S_{tot}, the K_M-values can be replaced by

$$K_i = S_{tot} \frac{1 - sat_{\nu_i}}{sat_{\nu_i}}, \ i = 1, 2.$$

This leads to an expression for $(V_1/V_2)_a$ which is independent from the total protein concentration but where the saturation of each reaction enters:

$$\begin{aligned}
\left(\frac{V_1}{V_2}\right)_a &= \frac{1 - a}{a} \frac{aS_{tot} + S_{tot}\frac{1 - sat_{\nu_1}}{sat_{\nu_1}}}{S_{tot} - aS_{tot} + S_{tot}\frac{1 - sat_{\nu_2}}{sat_{\nu_2}}} \\
&= \frac{1 - a}{a} \frac{sat_{\nu_2}}{sat_{\nu_1}} \frac{asat_{\nu_1} + 1 - sat_{\nu_1}}{(1 - a)sat_{\nu_2} + 1 - sat_{\nu_2}} = \frac{1 - a}{a} \frac{sat_{\nu_2}}{sat_{\nu_1}} \frac{1 + (a - 1)sat_{\nu_1}}{1 - asat_{\nu_2}}.
\end{aligned}$$

Forming the ratio of the velocity ratios, the following expression is derived for the response coefficient R_ν which signifies the steady state sensitivity of S_1 with respect to changes in V_1/V_2 in dependence on the saturation of the reactions:

$$R_\nu = \frac{(V_1/V_2)_{0.1}}{(V_1/V_2)_{0.9}} = \frac{\frac{1-0.1}{0.1}\frac{sat_{\nu_2}}{sat_{\nu_1}}\frac{1+(0.1-1)sat_{\nu_1}}{1-0.1sat_{\nu_2}}}{\frac{1-0.9}{0.9}\frac{sat_{\nu_2}}{sat_{\nu_1}}\frac{1+(0.9-1)sat_{\nu_1}}{1-0.9sat_{\nu_2}}} = \frac{9\frac{1-0.9sat_{\nu_1}}{1-0.1sat_{\nu_2}}}{\frac{1}{9}\frac{1-0.1sat_{\nu_1}}{1-0.9sat_{\nu_2}}}$$

$$= 81\frac{(1-0.9sat_{\nu_1})(1-0.9sat_{\nu_2})}{(1-0.1sat_{\nu_1})(1-0.1sat_{\nu_2})}.$$

On the basis of this equation, the response coefficients in Figure 6.5 C are calculated.

Derivation of R_{ν_1} and R_{ν_2} for the open switch

Introducing a constant production for S_1 and a degradation of S_2 governed by mass action kinetics into the system, the ODE system of the open switch reads as follows:

$$\begin{aligned}
\frac{dS_1}{dt} &= k_0 - V_1\frac{S_1}{S_1+K_1} + V_2\frac{S_2}{S_2+K_2}\\
\frac{dS_2}{dt} &= V_1\frac{S_1}{S_1+K_1} - V_2\frac{S_2}{S_2+K_2} - k_3S_2.
\end{aligned} \tag{A.5}$$

The value for S_2 at steady state is directly obtained by employing the result of the steady state equation for S_1

$$k_0 = V_1\frac{S_1^0}{S_1^0+K_1} - V_2\frac{S_2^0}{S_2^0+K_2}$$

for the steady state equation for S_2:

$$0 = k_0 - k_3S_2^0 \Leftrightarrow S_2^0 = \frac{k_0}{k_3}.$$

Hence, S_2^0 is independent from the conversion cycle reactions ν_1 and ν_2. Contrariwise, S_1^0 depends on all parameters of the system (k_3 entering via the concentration of S_2^0) where k_3 and K_2 can be lumped together in the saturation rate $sat_{\nu_2} = \frac{S_2^0}{S_2^0+K_2}$:

$$S_1^0 = K_1 \frac{\frac{k_0}{V_2} + \frac{S_2^0}{S_2^0+K_2}}{\frac{V_1}{V_2} - \frac{k_0}{V_2} - \frac{S_2^0}{S_2^0+K_2}} = K_1 \frac{\frac{k_0}{V_2} + sat_{\nu_2}}{\frac{V_1}{V_2} - \left(\frac{k_0}{V_2} + sat_{\nu_2}\right)} = K_1 \frac{k_0 + V_2 sat_{\nu_2}}{V_1 - (k_0 + V_2 sat_{\nu_2})}. \quad \text{(A.6)}$$

From Equation A.6 it is clear that the steady state of S_1 does not only depend on V_1/V_2, but both rate coefficients can occur separately. Thus, calculation of the response coefficient only towards changes in V_1/V_2 is not appropriate. Instead, the response coefficient with respect to changes in V_1, R_{ν_1}, and the response coefficient with respect to changes in V_2, R_{ν_2}, have to be considered.

Starting with the calculation of R_{ν_1}. Again, for the calculation of the response coefficient, the steady state is not necessary. Instead, the equation for V_1 in dependence on the steady state concentrations is derived. Directly from the ODE system (Equations A.5) and using $sat_{\nu_2} = \frac{S_2^0}{S_2^0+K_2}$, it follows

$$V_1 \frac{S_1^0}{S_1^0 + K_1} = k_0 + V_2 \frac{S_2^0}{S_2^0 + K_2}$$

$$\Leftrightarrow \qquad V_1 \frac{S_1^0}{S_1^0 + K_1} = k_0 + V_2 \cdot sat_{\nu_2}$$

$$\Leftrightarrow \qquad V_1 = \frac{(k_0 + V_2 \cdot sat_{\nu_2})(S_1^0 + K_1)}{S_1^0}.$$

The reaction velocities $(V_1)_a$ determines the value of V_1 necessary to have a steady state concentration for S_1 of aS_{thres}. Thus, it is

$$(V_1)_a = \frac{(k_0 + V_2 \cdot sat_{\nu_2})(aS_{thres} + K_1)}{aS_{thres}}.$$

Having a threshold value defined, the maximal saturation index obtained is the one for this threshold, i.e. $sat_{\nu_1} = \frac{S_{thres}}{S_{thres}+K_1}$. Hence, K_1 can be expressed by $K_1 = S_{thres}\frac{1-sat_{\nu_1}}{sat_{\nu_1}}$ leading to

$$(V_1)_a = \frac{(k_0 + V_2 \cdot sat_{\nu_2})(aS_{thres} + S_{thres}\frac{1-sat_{\nu_1}}{sat_{\nu_1}})}{aS_{thres}}$$

$$= (k_0 + V_2 \cdot sat_{\nu_2}) \cdot \left(1 + \frac{1-sat_{\nu_1}}{sat_{\nu_1}a}\right)$$

$$= (k_0 + V_2 \cdot sat_{\nu_2}) \cdot \frac{asat_{\nu_1} + 1 - sat_{\nu_1}}{sat_{\nu_1} a}.$$

For the response coefficient R_{ν_1}, one obtains

$$R_{\nu_1} = \frac{(V_1)_{0.1}}{(V_1)_{0.9}} = \frac{k_0 + V_2 \cdot sat_{\nu_2}}{k_0 + V_2 \cdot sat_{\nu_2}} \frac{(0.1 sat_{\nu_1} + 1 - sat_{\nu_1}) \cdot (sat_{\nu_1} 0.9)}{(0.9 sat_{\nu_1} + 1 - sat_{\nu_1}) \cdot (sat_{\nu_1} 0.1)}$$
$$= 9 \frac{1 - 0.9 sat_{\nu_1}}{1 - 0.1 sat_{\nu_1}}.$$

This relation is used for generating Figure 6.6 C.

The calculation of R_{ν_2} is shown in the following. First, V_2 needs to be expressed in dependence on S_1^0 and the other parameters. This is done starting from the ODE system (Equations A.5) and using $sat_{\nu_2} = \frac{S_2^0}{S_2^0 + K_2}$:

$$V_1 \frac{S_1^0}{S_1^0 + K_1} = k_0 + V_2 \frac{S_2^0}{S_2^0 + K_2}$$
$$\Leftrightarrow \qquad V_1 \frac{S_1^0}{S_1^0 + K_1} = k_0 + V_2 \cdot sat_{\nu_2}$$
$$\Leftrightarrow \qquad V_2 = \frac{V_1}{sat_{\nu_2}} \frac{S_1^0}{S_1^0 + K_1} - \frac{k_0}{sat_{\nu_2}}.$$

Note that in order to obtain positive values for V_2, it is required that $V_1 \frac{S_1^0}{S_1^0 + K_1} > k_0$. Furthermore, from Equation A.6 for the steady state of S_1 it follows that for altering V_2 from zero to infinity, S_1^0 can be altered only in the open interval $[S_{min}, +\infty)$, with $S_{min} = K_1 \frac{k_0}{V_1 - k_0}$. Thus, if considering a maximal threshold concentration S_{thres}, for example, 10% of the maximal response are not given by $0.1 \cdot S_{thres}$, but rather by $S_{min} + (S_{thres} - S_{min}) \cdot 0.1$. Consequently, also the $(V_2)_a$ are accordingly defined. Note that $S_{min} > 0$ requires $V_1 > k_0$, which is assumed in the following and leads in particular also to positive values for V_2 in dependence on S_1^0. Additionally, since $K_1 = S_{thres} \frac{1 - sat_{\nu_1}}{sat_{\nu_1}}$, in order to obtain $S_{min} < S_{thres}$ and thus to be able to define the response coefficient for any threshold S_{thres}, it is required that

$$K_1 \frac{k_0}{V_1 - k_0} < K_1 \frac{sat_{\nu_1}}{1 - sat_{\nu_1}}$$
$$\Leftrightarrow \qquad \frac{k_0}{V_1 - k_0} < \frac{sat_{\nu_1}}{1 - sat_{\nu_1}}.$$

For the following derivation of $(V_2)_a$, it is employed that $S_{min} = K_1 \frac{k_0}{V_1 - k_0}$ and furthermore that $K_1 = S_{thres} \frac{1 - sat_{\nu_1}}{sat_{\nu_1}}$.

$$
\begin{aligned}
(V_2)_a &= \frac{V_1}{sat_{\nu_2}} \frac{S_{min} + (S_{thres} - S_{min}) \cdot a}{S_{min} + (S_{thres} - S_{min}) \cdot a + K_1} - \frac{k_0}{sat_{\nu_2}} \\[2mm]
&= \frac{V_1}{sat_{\nu_2}} \frac{K_1 \frac{k_0}{V_1 - k_0} + (S_{thres} - K_1 \frac{k_0}{V_1 - k_0}) \cdot a}{K_1 \frac{k_0}{V_1 - k_0} + (S_{thres} - K_1 \frac{k_0}{V_1 - k_0}) \cdot a + K_1} - \frac{k_0}{sat_{\nu_2}} \\[2mm]
&= \frac{V_1}{sat_{\nu_2}} \frac{\frac{k_0}{V_1 - k_0} + (\frac{S_{thres}}{K_1} - \frac{k_0}{V_1 - k_0}) \cdot a}{\frac{k_0}{V_1 - k_0} + (\frac{S_{thres}}{K_1} - \frac{k_0}{V_1 - k_0}) \cdot a + 1} - \frac{k_0}{sat_{\nu_2}} \\[2mm]
&= \frac{1}{sat_{\nu_2}} \frac{(V_1 - k_0)(\frac{k_0}{V_1 - k_0} + (\frac{S_{thres}}{K_1} - \frac{k_0}{V_1 - k_0}) \cdot a) - k_0}{\frac{k_0}{V_1 - k_0} + (\frac{S_{thres}}{K_1} - \frac{k_0}{V_1 - k_0}) \cdot a + 1} \\[2mm]
&= \frac{1}{sat_{\nu_2}} \frac{(k_0 + (\frac{S_{thres}(V_1 - k_0)}{K_1} - k_0) \cdot a) - k_0}{\frac{k_0}{V_1 - k_0} + (\frac{S_{thres}}{K_1} - \frac{k_0}{V_1 - k_0}) \cdot a + 1} \\[2mm]
&= \frac{1}{sat_{\nu_2}} \frac{(\frac{S_{thres}(V_1 - k_0)}{K_1} - k_0) \cdot a}{\frac{k_0}{V_1 - k_0} + (\frac{S_{thres}}{K_1} - \frac{k_0}{V_1 - k_0}) \cdot a + 1} \\[2mm]
&= \frac{1}{sat_{\nu_2}} \frac{(\frac{sat_{\nu_1}(V_1 - k_0)}{1 - sat_{\nu_1}} - k_0) \cdot a}{\frac{k_0}{V_1 - k_0} + (\frac{sat_{\nu_1}}{1 - sat_{\nu_1}} - \frac{k_0}{V_1 - k_0}) \cdot a + 1}.
\end{aligned}
$$

Consequently, for the response coefficient R_{ν_2}, it is obtained

$$
\begin{aligned}
R_{\nu_2} = \frac{(V_2)_{0.9}}{(V_2)_{0.1}} &= \frac{sat_{\nu_2}}{sat_{\nu_2}} \frac{((\frac{sat_{\nu_1}(V_1 - k_0)}{1 - sat_{\nu_1}} - k_0) \cdot 0.9) \cdot (\frac{sat_{\nu_1}}{1 - sat_{\nu_1}} - \frac{k_0}{V_1 - k_0}) \cdot 0.1 + 1)}{(\frac{sat_{\nu_1}(V_1 - k_0)}{1 - sat_{\nu_1}} - k_0) \cdot 0.1) \cdot (\frac{sat_{\nu_1}}{1 - sat_{\nu_1}} - \frac{k_0}{V_1 - k_0}) \cdot 0.9 + 1)} \\[2mm]
&= 9 \frac{\frac{sat_{\nu_1}}{1 - sat_{\nu_1}} - \frac{k_0}{V_1 - k_0}) \cdot 0.1 + 1}{\frac{sat_{\nu_1}}{1 - sat_{\nu_1}} - \frac{k_0}{V_1 - k_0}) \cdot 0.9 + 1}.
\end{aligned}
$$

This relation is used for generating Figure 6.6 E.

Derivation of R_ν for the short chain

Removing the backward reaction ν_2 from the open switch, the following ODE system for the short chain is obtained:

$$\begin{aligned}
\frac{dS_1}{dt} &= k_0 - V_1 \frac{S_1}{S_1 + K_1} \\
\frac{dS_2}{dt} &= V_1 \frac{S_1}{S_1 + K_1} - k_2 S_2.
\end{aligned} \tag{A.7}$$

Again, using the steady state equation for S_1 and the steady state equation for S_2, it is $0 = k_0 - k_2 S_2^0$ yielding $S_2^0 = k_0/k_2$. S_2^0 is again independent from the conversion reaction ν_1. The steady state of S_1 is derived by straightforward calculation:

$$\begin{aligned}
& V_1 \frac{S_1^0}{S_1^0 + K_1} = k_0 \\
\Leftrightarrow \quad & V_1 S_1^0 = k_0 S_1^0 + k_0 K_1 \\
\Leftrightarrow \quad & S_1^0 = \frac{k_0 K_1}{V_1 - k_0}.
\end{aligned}$$

It is independent from the degradation reaction ν_2, but depends on the conversion reaction ν_1 and the production rate coefficient k_0. Hence, only the steady state concentration change of S_1 can be examined with respect to the remaining conversion rate coefficient V_1.

In order to calculate the according response coefficient, not the steady state concentration of S_1 is important but the dependence of V_1 on the steady state concentration of S_1 and the other parameters:

$$V_1 = k_0 \frac{S_1^0 + K_1}{S_1^0}.$$

Setting a threshold value of S_{thres}, the saturation of reaction ν_1 is given by

$$sat_{\nu_1} = \frac{S_{thres}}{S_{thres} + K_1},$$

K_1 can be therefore replaced by $K_1 = \frac{1-sat_{\nu_1}}{sat_{\nu_1}} S_{thres}$ in the equation for V_1

$$V_1 = \frac{k_0}{S_1^0}(S_1^0 + \frac{1 - sat_{\nu_1}}{sat_{\nu_1}} S_{thres}).$$

Let $(V_1)_a$ be the rate coefficient V_1 necessary to obtain a steady state concentration of S_1 of aS_{thres}. It follows

$$(V_1)_a = k_0 \frac{aS_{thres} + \frac{1-sat_{\nu_1}}{sat_{\nu_1}} S_{thres}}{aS_{thres}} = k_0 \frac{asat_{\nu_1} + 1 - sat_{\nu_1}}{asat_{\nu_1}} = k_0 \frac{1 - (1-a)sat_{\nu_1}}{asat_{\nu_1}}.$$

The response coefficient of the steady state of S_1 with respect to changes in V_1 is hence given by

$$R_\nu = \frac{(V_1)_{0.1}}{(V_1)_{0.9}} = \frac{k_0 \frac{(1-(1-0.1)sat_{\nu_1})}{0.1sat_{\nu_1}}}{k_0 \frac{(1-(1-0.9)sat_{\nu_1})}{0.9sat_{\nu_1}}} = \frac{(1 - 0.9sat_{\nu_1})0.9sat_{\nu_1}}{(1 - 0.1sat_{\nu_1})0.1sat_{\nu_1}} = 9\frac{1 - 0.9sat_{\nu_1}}{1 - 0.1sat_{\nu_1}}.$$

This relation is used for the generation of Figure 6.7 C.

A.4.3. Steady state response coefficients according to MCA

Response coefficients $R_{V_1/V_2}^{S_1^0}$ for the Goldbeter-Koshland switch

Analytically deriving the response coefficients $R_{V_1/V_2}^{S_1^0}$ for the Goldbeter-Koshland switch is possible, but it requires a lengthy calculation and therefore only numerical results for the maximal response coefficients of S_1^0 obtained for varying V_1/V_2, while keeping the other parameters fixed, are given in dependence on sat_{ν_1} and sat_{ν_2} in Figure A.18 B.

Similar to the response coefficient R_ν from section 6.2.3, the maximal response coefficient increases for increasing saturation of reactions ν_1 and ν_2. However, it is not symmetrical with respect to the saturation levels of the two reactions. This is due to the relative change of S_1 being locally not the same as the relative change of $S_2^0 = 1 - S_1^0$. For example, if S_1^0 changes by 20%, i.e. $\overline{S_1^0} = 1.2 \cdot S_1^0$, this coincides with a change in S_2^0 of $\overline{S_2^0} = 1 - 1.2 \cdot S_1^0$, which does not represent a change of 20% in S_2^0 in general (unless $S_1^0 = S_2^0 = 0.5$). Thus, in contrast to the response

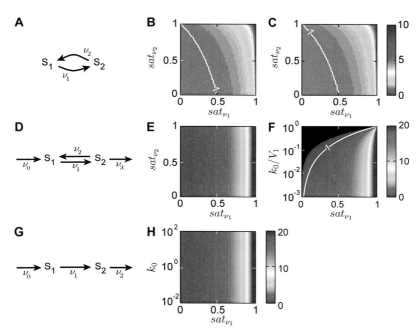

Figure A.18.: Maximal absolute steady state response coefficients according to MCA. White level lines indicate absolute response coefficient values of 1 (linear sensitivity). A: Goldbeter-Koshland switch. B, C: Maximal absolute response coefficients of the steady state of S_1 for the Goldbeter-Koshland switch for changes in V_1/V_2, $max(|R^{S_1^0}_{V_1/V_2}|)$ (A), or V_2/V_1, $max(|R^{S_1^0}_{V_2/V_1}|)$ (B). D: Open switch. E: Maximal absolute response coefficients for $S_1^0 \in [0, S_{thres}]$ for the open switch for changes in V_1, $max(|R^{S_1^0}_{V_1}|)$, according to Equation A.8 in dependence on sat_{ν_1} and sat_{ν_2}. F: Maximal absolute response coefficients for $S_1^0 \in [S_{min}, S_{thres}]$ for the open switch for changes in V_2, $max(|R^{S_1^0}_{V_2}|)$, according to Equation A.9 in dependence on sat_{ν_1} and k_0/V_1. Regions where the response coefficient is not defined are marked in black. G: Short chain. H: Maximal absolute response coefficients for $S_1^0 \in [0, S_{thres}]$ for the short chain for changes in V_1, $max(|R^{S_1^0}_{V_1}|)$, according to Equation A.10 in dependence on sat_{ν_1} and k_0.

coefficient R_ν shown in section 6.2.3 which is the same for changes in S_1^0 and changes in S_2^0, the response coefficients $R_{V_1/V_2}^{S_i^0}$ for S_1^0 and S_2^0 are not identical. In particular, the response coefficient needs not to be symmetrical in sat_{ν_1} and sat_{ν_2}. However, since the Goldbeter-Koshland switch is symmetric, the response coefficient for S_2^0 for perturbations in V_2/V_1, $R_{V_2/V_1}^{S_2^0}$, is the same as the response coefficient for S_1^0 for perturbations in V_1/V_2, $R_{V_1/V_2}^{S_1^0}$, for interchanging sat_{ν_1} and sat_{ν_2}.

The response coefficients for changes in V_1/V_2 and changes in V_2/V_1 remain identical up to the sign. In particular, using $f(x) = 1/x$ and the rule for deriving concatenated functions, it is

$$
\begin{aligned}
R_{V_2/V_1}^{S} &= R_{f(V_1/V_2)}^{S} = \frac{f(V_1/V_2)}{S} \frac{\partial S}{\partial f(V_1/V_2)} = \frac{f(V_1/V_2)}{S} \frac{\partial S}{\partial V_1/V_2} \Big/ \frac{\partial f(V_1/V_2)}{\partial V_1/V_2} \\
&= \frac{V_2/V_1}{S} \frac{\partial S}{\partial V_1/V_2} \Big/ (-(V_2/V_1)^2) = -\frac{V_1/V_2}{S} \frac{\partial S}{\partial V_1/V_2} = -R_{V_1/V_2}^{S}.
\end{aligned}
$$

Consequently, the results for the absolute maximal response coefficient for changes in S_1^0 with regard to changes in V_2/V_1 are the same as for changes in V_1/V_2 (compare Figure A.18 B to C).

Response coefficients $R_{V_1}^{S_1^0}$ and $R_{V_2}^{S_1^0}$ for the open switch

As derived in section A.4.2, for the open switch, the steady state concentration of $S_2 = k_0/k_2$ is independent from the rate coefficients V_1 and V_2. Consequently, define $sat_{\nu_2} = S_2^0/(S_2^0 + K_2)$. For S_1^0, it is then

$$
S_1^0 = K_1 \frac{k_0 + sat_{\nu_2} V_2}{V_1 - (k_0 + sat_{\nu_2} V_2)}.
$$

The response coefficient of S_1^0 with respect to changes in V_1, $R_{V_1}^{S_1^0}$, is given by

$$
R_{V_1}^{S_1^0} = \frac{\partial S_1^0}{\partial V_1} \frac{V_1}{S_1^0} = -K_1 \frac{k_0 + sat_{\nu_2} V_2}{(V_1 - (k_0 + sat_{\nu_2} V_2))^2} \frac{V_1}{S_1^0} = -\frac{V_1}{(V_1 - (k_0 + sat_{\nu_2} V_2))}.
$$

Replacing $V_1 = (k_0 + sat_{\nu_2} V_2) \frac{S_1^0 + K_1}{S_1^0}$ in order to be able to maximize the response coefficient $R_{V_1}^{S_1^0}$ over a range of steady state concentrations up to a threshold value

S_{thres}, i.e. $S_1^0 \in [0, S_{thres}]$, one gets

$$R_{V_1}^{S_1^0} = -\frac{(k_0 + sat_{\nu_2}V_2)\frac{S_1^0+K_1}{S_1^0}}{(k_0 + sat_{\nu_2}V_2)(\frac{S_1^0+K_1}{S_1^0} - 1)} = -\frac{S_1^0 + K_1}{K_1}.$$

If using $sat_{\nu_1} = \frac{S_{thres}}{S_{thres}+K_1}$, which is the general saturation index, i.e. the maximal saturation index obtained in a curve for $S_1^0 \leq S_{thres}$ in analogy to the saturation index employed for the response coefficients R_{ν_1}, K_1 can be replaced by $K_1 = S_{thres}\frac{1-sat_{\nu_1}}{sat_{\nu_1}}$. The maximal response coefficient which could be obtained for $S_1^0 \in [0, S_{thres}]$ is thus

$$max_{S_1^0 \in [0,S_{thres}]}(|R_{V_1}^{S_1^0}|) = \frac{S_{thres} + S_{thres}\frac{1-sat_{\nu_1}}{sat_{\nu_1}}}{S_{thres}\frac{1-sat_{\nu_1}}{sat_{\nu_1}}} = \frac{1 + \frac{1-sat_{\nu_1}}{sat_{\nu_1}}}{\frac{1-sat_{\nu_1}}{sat_{\nu_1}}} = \frac{1}{1 - sat_{\nu_1}}. \quad \text{(A.8)}$$

In other words, the maximal absolute response coefficient $|R_{V_1}^{S_1^0}|$ depends exclusively on the saturation level of ν_1, exactly as the response coefficient R_{ν_1} shown in section 6.2.4. Depicting $max(|R_{V_1}^{S_1^0}|)$ in dependence on sat_{ν_1} and sat_{ν_2} according to Equation A.8 again emphasizes its independence from sat_{ν_2} (Figure A.18 E). Similar to the response coefficient R_{ν_1} of the open switch (section 6.2.4), the steady state S_1^0 is found to be always ultrasensitive with respect to changes in V_1 since $max(|R_{V_1}^{S_1^0}|)$ takes only values above 1. Additionally, in accordance with results from R_{ν_1}, S_1^0 can be deliberately sensitive to changes in V_1 for approaching complete saturation of ν_1 according to $max(|R_{V_1}^{S_1^0}|)$ since $max(|R_{V_1}^{S_1^0}|) \to \infty$ for $sat_{\nu_1} \to 1$.

For the response coefficient towards changes in V_2, it is

$$R_{V_2}^{S_1^0} = \frac{\partial S_1^0}{\partial V_2}\frac{V_1}{S_1^0} = K_1 \frac{sat_{\nu_2} \cdot (V_1 - (k_0 + sat_{\nu_2}V_2)) + (k_0 + sat_{\nu_2}V_2) \cdot sat_{\nu_2}}{(V_1 - (k_0 + sat_{\nu_2}V_2))^2} \frac{V_1}{S_1^0}$$

$$= \frac{sat_{\nu_2} \cdot V_2 \cdot V_1}{(V_1 - (k_0 + sat_{\nu_2}V_2))(k_0 + sat_{\nu_2}V_2)}$$

Replacing $V_2 = \frac{V_1}{sat_{\nu_2}}\frac{S_1^0}{S_1^0+K_1} - \frac{k_0}{sat_{\nu_2}}$ in order to be able to maximize the absolute response coefficient over a range of concentrations, $S_1^0 \in [S_{min}, S_{thres}]$ (note that for varying V_2, S_1^0 cannot take concentrations arbitrarily close to 0), one obtains

$$R_{V_2}^{S_1^0} = \frac{(V_1 \frac{S_1^0}{S_1^0+K_1} - k_0) \cdot V_1}{(V_1 - (k_0 + (V_1 \frac{S_1^0}{S_1^0+K_1} - k_0)))(k_0 + (V_1 \frac{S_1^0}{S_1^0+K_1} - k_0))}$$

$$= \frac{(V_1 \frac{S_1^0}{S_1^0+K_1} - k_0) \cdot V_1}{(V_1 - (V_1 \frac{S_1^0}{S_1^0+K_1}))(V_1 \frac{S_1^0}{S_1^0+K_1})} = \frac{(V_1 \frac{S_1^0}{S_1^0+K_1} - k_0)}{(1 - \frac{S_1^0}{S_1^0+K_1})(V_1 \frac{S_1^0}{S_1^0+K_1})}$$

$$= \frac{(V_1 \frac{S_1^0}{S_1^0+K_1} - k_0)}{\frac{K_1}{S_1^0+K_1}(V_1 \frac{S_1^0}{S_1^0+K_1})} = \frac{(V_1 \cdot S_1^0 - k_0(S_1^0 + K_1))(S_1^0 + K_1)}{K_1 \cdot V_1 \cdot S_1^0}$$

$$= \frac{(S_1^0 - \frac{k_0}{V_1}(S_1^0 + K_1))(S_1^0 + K_1)}{K_1 \cdot S_1^0}.$$

Using $K_1 = S_{thres} \frac{1-sat_{\nu_1}}{sat_{\nu_1}}$ for the general saturation index sat_{ν_1} yields

$$R_{V_2}^{S_1^0} = \frac{(S_1^0 - \frac{k_0}{V_1}(S_1^0 + S_{thres}\frac{1-sat_{\nu_1}}{sat_{\nu_1}}))(S_1^0 + S_{thres}\frac{1-sat_{\nu_1}}{sat_{\nu_1}})}{S_{thres}\frac{1-sat_{\nu_1}}{sat_{\nu_1}} \cdot S_1^0}$$

$$= \frac{(S_1^0 - \frac{k_0}{V_1}(S_1^0 + S_{thres}(1 - sat_{\nu_1})))(S_1^0 + S_{thres}(1 - sat_{\nu_1}))}{S_{thres}(1 - sat_{\nu_1}) \cdot S_1^0 \cdot sat_{\nu_1}}$$

Note that the response coefficient is only defined for $S_{min} \leq S_{thres}$. Thus, it is required that $K_1 \frac{k0/V_1}{1-k_0/V_1} \leq S_{thres}$. Using $K_1 = S_{thres}\frac{1-sat_{\nu_1}}{sat_{\nu_1}}$, straightforward calculation reveals that $S_{min} \leq S_{thres}$ and thus the definition of $R_{V_2}^{S_1^0}$ in dependence on S_{thres} requires $k_0/V_1 < sat_{\nu_1}$. In order to determine the maximal value of $|R_{V_1}^{S_1^0}|$ for $S_1^0 \in [S_{min}, S_{thres}]$, zeros of the first derivative of $R_{V_2}^{S_1^0}$ with respect to S_1^0 are calculated (detailed calculation not shown):

$$(S_1^0)_{1/2} = \pm S_{thres} \frac{1 - sat_{\nu_1}}{sat_{\nu_1}} \sqrt{\frac{k_0}{k_0 - V_1}}.$$

Since $k_0 < V_1$ in order to obtain $S_1^0 > 0$, these zeros are never real values. Consequently, there are no minima or maxima for positive steady state concentrations of S_1, and thus the maximal values of $|R_{V_2}^{S_1^0}|$ are obtained at the borders of the permitted interval $[S_{min}, S_{thres}]$. Indeed, the maximal value is taken at S_{thres} (not shown), and where defined, the value of the maximal response coefficient does not depend on S_{thres}:

$$max(|R_{V_2}^{S_1^0}|) = \frac{sat_{\nu_1} - \frac{k_0}{V_1}}{sat_{\nu_1}(1 - sat_{\nu_1})}. \tag{A.9}$$

From Equation A.9 follows that $max(|R_{V_2}^{S_1^0}|)$ depends only on sat_{ν_1} and on k_0/V_1. The values for the maximal response coefficient of S_1^0 by changes of V_2 are given in dependence on these two entities in Figure A.18 F. Note that S_1^0 is more sensitive to changes in V_2 the more ν_1 is saturated and the smaller k_0/V_1 is, which is exactly the same tendency as observed for the response coefficient R_{ν_2} in section 6.2.4.

Response coefficients $R_{V_1}^{S_1^0}$ for the short chain

For the short chain model, the steady state of $S_2 = k_0/k_2$ is constant for varying V_1. It is derived in section A.4.2 that the steady state concentration of S_1 is given by

$$S_1^0 = \frac{k_0 K_1}{V_1 - k_0}.$$

The response coefficient $R_{V_1}^{S_1^0}$ can be straightforwardly calculated via

$$R_{V_1}^{S_1^0} = \frac{\partial S_1^0}{\partial V_1} \frac{V_1}{S_1^0} = -K_1 \frac{k_0}{(V_1 - k_0)^2} \frac{V_1}{S_1^0} = -\frac{V_1}{V_1 - k_0}.$$

Using $V_1 = \frac{k_0(S_1^0 + K_1)}{S_1^0}$ and, for a fixed threshold concentration S_{thres} and examining the response coefficients for $S_1 \in [0, S_{thres}]$, $K_1 = S_{thres}\frac{1 - sat_{\nu_1}}{sat_{\nu_1}}$ for the general saturation index $sat_{\nu_1} = max_{S_1^0 \in [0, S_{thres}]}(\frac{S_1^0}{S_1^0 + K_1} = \frac{S_{thres}}{S_{thres} + K_1})$, it is obtained

$$R_{V_1}^{S_1^0} = -\frac{\frac{k_0(S_1^0 + K_1)}{S_1^0}}{\frac{k_0(S_1^0 + K_1)}{S_1^0} - k_0} = -\frac{\frac{k_0(S_1^0 + K_1)}{S_1^0}}{k_0(\frac{S_1^0 + K_1}{S_1^0} - 1)} = -\frac{\frac{S_1^0 + K_1}{S_1^0}}{(\frac{K_1}{S_1^0})}$$

$$= -\frac{S_1^0 + K_1}{K_1} = -\frac{S_1^0 + S_{thres}\frac{1 - sat_{\nu_1}}{sat_{\nu_1}}}{S_{thres}\frac{1 - sat_{\nu_1}}{sat_{\nu_1}}}$$

Consequently, the maximal absolute response coefficient for $S_1^0 \in [0, S_{thres}]$ is

$$
\begin{aligned}
max_{S_1^0 \in [0, S_{thres}]}(|R_{V_1}^{S_1^0}|) &= \frac{S_{thres} + S_{thres}\frac{1-sat_{\nu_1}}{sat_{\nu_1}}}{S_{thres}\frac{1-sat_{\nu_1}}{sat_{\nu_1}}} \\
&= \frac{S_{thres}(sat_{\nu_1} + (1 - sat_{\nu_1}))}{S_{thres}(1 - sat_{\nu_1})} = \frac{1}{1 - sat_{\nu_1}}.
\end{aligned}
\tag{A.10}
$$

This is exactly the same expression as for the steady state sensitivity of S_1 for changes in V_1 for the open switch. The values for $max|R_{V_1}^{S_1^0}|$ are depicted in Figure A.18 H. Since $max|R_{V_1}^{S_1^0}|$ depends only on sat_{ν_1}, yields values between linear sensitivity $(max|R_{V_1}^{S_1^0}| \to 1$ for $sat_{\nu_1} \to 0)$ and infinite ultrasensitivity $(max|R_{V_1}^{S_1^0}| \to \infty$ for $sat_{\nu_1} \to 1)$, the maximal absolute response coefficient $max|R_{V_1}^{S_1^0}|$ reveals exactly the same conclusions on the steady state sensitivity as the response coefficient R_ν shown in section 6.2.5.

A.4.4. Stability of the chain models with saturating kinetics

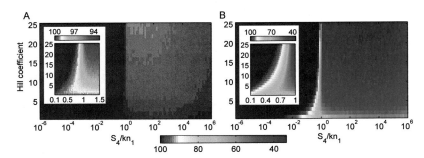

Figure A.19.: Stability of the chain models with saturating kinetics. S_4/kn_1 determines the size of the feedback term in case of negative and positive feedback. $1 \cdot 10^4$ parameter sets are examined for each combination of an integer Hill coefficient in $[1, 25]$ and each feedback-to-constant-ratio in one of the intervals $\{(10^{-6}, 10^{-5.9}), \ldots, (10^{5.9}, 10^6)\}$ for the chain model with either negative (A) or positive feedback (B), and the eigenvalues of the Jacobian matrix are determined. Indicated is the percentage of stable sets meaning that all real parts of the eigenvalues are negative. The insets show the details for values of S_4/kn_1 close to 1.

Via the Michaelis-Menten kinetics, a new source of nonlinearity is introduced in the models. In Figure A.19, the resulting new distribution of stability of the models in dependence on the Hill coefficient of the feedback term is given. The chain models with saturating kinetics and negative (panel A) or positive feedback (panel B) show unstable steady states already for lower Hill coefficients than for the model with mass action kinetics (section 4.2.2). A Hill coefficient of 1 is sufficient for both feedback types to yield oscillations which complies with very low cooperativity.

A.4.5. Hill coefficients in the chain models with saturating kinetics

The chain models with saturating kinetics also allow oscillations for a Hill coefficient of 1 (section A.4.4). The sensitivities of the according chain models are analyzed and compared to the sensitivities obtained for the models with Hill coefficients of $n = 9$ (negative feedback model) or $n = 2$ (positive feedback model) in Figure A.20. Decreasing the Hill coefficient to $n = 1$ increases both amplitude and period sensitivities. Hence, the effect of increased sensitivities for the models with saturating kinetics compared to the models with linear mass action kinetics is even enhanced if additionally altering the Hill coefficient to the lowest possible integer value.

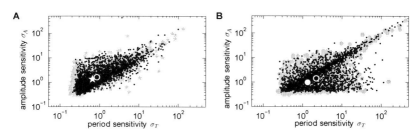

Figure A.20.: Influence of the Hill coefficient in the chain models with saturating kinetics. A (B): The sensitivities of the original chain model with negative feedback (positive feedback) and a Hill coefficient of $n = 9$ ($n = 2$) in gray, white symbol for median values, is compared to the sensitivities for the model with a Hill coefficient of $n = 1$ in black, black circle for median values.

A.4.6. Influence of the definition of a saturated reaction

The influence of the choice of the saturation border for the examination of the number of saturated Michaelis-Menten terms for the chain models with saturating is examined (Figure A.21 A, B for the negative feedback chain model, C, D for the positive feedback chain model). Basis of the data are the 2500 parameter sets found during the sensitivity analysis of the chain models with saturating kinetics. For each parameter, it is determined how many of the seven Michaelis-Menten-governed reaction rates are saturated (according to the definition given in the legend) and the median period sensitivities and median amplitude sensitivities are examined in

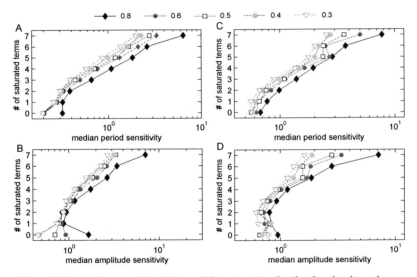

Figure A.21.: Influence of the choice of the saturation border for the dependency of the median period sensitivities (A, C) and median amplitude sensitivities (B, D) from the number of saturated reactions in the negative feedback chain model (A, B) or positive feedback chain model (C, D). A reaction ν_i is considered to be saturated for if its saturation index satisfies $sat_i > 0.3, 0.4, 0.5, 0.6$ or 0.8 according to the legend. For each of the 2500 parameter sets for each model, the numbers of saturated reactions is counted; for each category, the median for all sensitivities of the according parameter sets is indicated. Lines are included for better visibility.

each category. For both figures, in A and B, respectively, the dependency of median period sensitivity and of the median amplitude sensitivity, respectively, from the number of saturated terms are shown. The general trends of increase in median sensitivities for increase in number of saturated reactions holds also for the different definitions of a saturated reaction as examined here and hence does not depend on the particular choice of the value of the saturation border.

A.4.7. Particular reactions being saturated

Negative feedback chain models

The saturation of reactions ν_3, ν_4, ν_6 display the strongest influence on the period and amplitude sensitivities of the negative feedback chain model with saturating kinetics. Here, it is in more detail examined how the saturation of the other reactions influence the sensitivities if the saturation of reactions ν_3, ν_4, ν_6 are kept constant. Therefore, for each of the eight possible saturation combinations, the according parameter sets from the 2500 are examined. For each subset and each of the reactions ν_2, ν_5, ν_7, ν_8, the two groups of parameter sets are compared which yield either ν_i saturated (dark symbols in Figure A.22) or non-saturated (white symbols in Figure A.22). Significant differences between the sensitivities are marked by stars in Figure A.22 (MWU p-values not shown).

It is revealed that for constant saturation levels of ν_3, ν_4, ν_6, altering the saturation level of reaction ν_7 does not induce any significant changes in sensitivities. Altering the saturation of reaction ν_2 only delivers significant changes in the period sensitivities under only 2 of the 8 conditions (Figure A.22 A, G). Furthermore, for only ν_6 being saturated (Figure A.22 D), for ν_3 and ν_4 being saturated (Figure A.22 E) and for ν_3 and ν_4 being saturated (Figure A.22 F), alteration of the saturation of any individual of the reactions ν_2, ν_5, ν_7, ν_8 does not influence the sensitivities significantly.

This indicates that the saturation levels of ν_2 and ν_7 are far less important for the sensitivities than those of reactions ν_5 and ν_8, and the saturation of all four reactions is affecting the sensitivities less than those of reactions ν_3, ν_4, ν_6 in the negative feedback chain model with saturating kinetics.

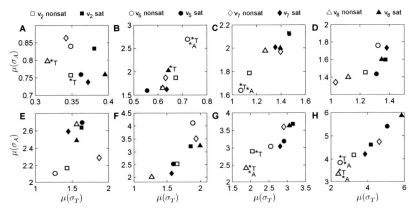

Figure A.22.: Effect of the saturations of reactions other than ν_3, ν_4, ν_6 in the negative feedback chain model. The saturation of ν_3, ν_4, ν_6 are kept fixed and the median period and amplitude sensitivities for one of the other reactions ν_2, ν_5, ν_7, ν_8 being saturated (black symbols) or not saturated (white symbols) are given. A: ν_3, ν_4, ν_6 not saturated (1133); B: ν_3 saturated, ν_4, ν_6 not saturated (228); C: ν_4 saturated, ν_3, ν_6 not saturated (485); D: ν_6 saturated, ν_3, ν_4 not saturated (285); E: ν_3, ν_4 saturated, ν_6 not saturated (126); F: ν_3, ν_6 saturated, ν_4 not saturated (70); G: ν_4, ν_6 saturated, ν_3 not saturated (132); H: ν_3, ν_4, ν_6 saturated (41). The number in brackets give the numbers of parameter sets in each group. Significant differences according to the MWU test are denoted by *T and *A for the period and amplitude sensitivities, respectively.

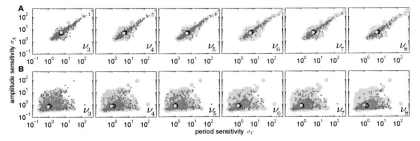

Figure A.23.: Effect of saturation of the other reactions if ν_2 is saturated (A) or non-saturated (B) in the positive feedback chain model with saturating kinetics. Dark gray: reaction indicated in the lower right is saturated, light gray: reaction is not saturated. Median values are indicated by circles in the according colors.

Table A.7.: Results for the medians and MWU test comparing the sensitivity distributions of the chain models for a reaction (rct) being non-saturated (nsat) to the reaction being saturated (sat) for the saturation of ν_2 being fixed, either saturated or not. Given are the medians (μ), the MWU p-values (MWU p), U, z-values (z-val) and the number of parameter sets obtained in each category (par. sets). Data corresponds to Figure A.23.

rct	ν_2	period sensitivity					amplitude sensitivity					par. sets	
		μ nsat	μ sat	MWU p	U	z-val	μ nsat	μ sat	MWU p	U	z-val	nsat	sat
ν_3	sat	5.59	5.51	0.42	-0.2	16024	5.72	4.97	0.40	-0.3	15950	81	401
	nsat	0.88	1.06	$5.5 \cdot 10^{-5}$	-3.9	391382	0.71	0.83	$2.1 \cdot 10^{-4}$	-3.5	395453	633	1385
ν_4	sat	5.15	6.06	0.0079	-2.4	25346	4.73	5.30	0.067	-1.5	26739	245	237
	nsat	0.89	1.22	$5.8 \cdot 10^{-10}$	-6.1	407502	0.78	0.82	0.59	0.2	488156	1227	791
ν_5	sat	5.10	5.70	0.18	-0.9	26583	4.75	5.14	0.25	-0.7	26916	194	288
	nsat	0.97	1.05	$8.3 \cdot 10^{-5}$	-3.8	459707	0.77	0.81	0.13	-1.1	494246	1001	1017
ν_6	sat	5.24	6.09	0.057	-1.6	24144	4.91	5.51	0.076	-1.4	24362	313	169
	nsat	0.94	1.12	$2.4 \cdot 10^{-5}$	-4.1	376434	0.79	0.80	0.38	-0.3	421249	1419	599
ν_7	sat	5.15	5.73	0.12	-1.2	27150	5.00	5.11	0.40	-0.3	28527	257	225
	nsat	0.98	1.07	0.0039	-2.7	450859	0.80	0.79	0.47	-0.1	484113	1229	789
ν_8	sat	5.30	6.31	0.17	-0.9	24864	5.05	5.05	0.70	0.5	27016	316	66
	nsat	0.90	1.28	$1.2 \cdot 10^{-16}$	-8.1	355206	0.83	0.72	$3.9 \cdot 10^{-7}$	-4.9	395634	1332	686

Positive feedback chain model

The saturation of reaction ν_2 has revealed to yield very strong influence on the period and amplitude sensitivity of the chain model with positive feedback. Now, the effect of the saturation of ν_2 is decoupled from the saturation of the other reactions. Therefore, the populations of the 482 parameter sets where ν_2 is saturated (upper panels) and the populations of the remaining 2018 parameter sets where ν_2 is not saturated (lower panels) are investigated (Figure A.23). For each reaction ν_i (given in the lower right of the panels in Figure A.23) other than ν_2 and each population, the sensitivities of the two subpopulations with the reaction being saturated (dark) or not saturated (light) are compared. The differences in the two subpopulations each are compared to the differences obtained for all 2500 parameter sets depicted in Figure 6.12 C in Table A.7.

In fact, for the population in which ν_2 is not saturated, altering the saturation level of any of the other reactions always induces a significant increase in the period sensitivities. For the amplitude sensitivities, this is only the case for ν_3 and ν_8.

The only significant difference for the population of ν_2 being saturated is found for altering the saturation of ν_4 which increases the period sensitivities. One can conclude that ν_2 yields indeed the strongest impact on the sensitivities for changing its saturation level, and that the system is decisively more susceptible to alterations in any of the other reactions saturation levels if ν_2 is not saturated.

A.4.8. Chain models only partly employing saturating kinetics

Here, the chain models are investigated yielding Michaelis-Menten kinetics only in their degradation reactions (ν_3, ν_5, ν_7, ν_8), only in their inner-loop conversion reactions (ν_4, ν_6) or only in their inner-loop conversions and ν_2. The sensitivities obtained for a sensitivity analysis are given in Figure A.24 A-D for the negative feedback chain model, Figure A.24 E-H for the positive feedback chain model. Os-

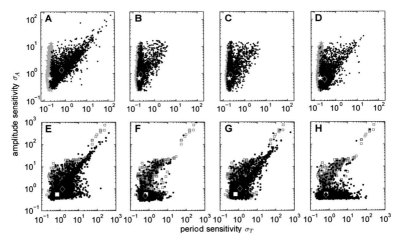

Figure A.24.: Effect of different reactions employing saturating kinetics for the negative feedback chain model (A-D) and the positive feedback chain model (E-H). A and E: Saturating kinetics in all seven reactions. B and F: Saturating kinetics in the inner-loop conversions ν_4 and ν_6. C and G: Saturating kinetics in all conversions (ν_2, ν_4, ν_6). D and H: Saturating kinetics in all degradations (ν_3, ν_5, ν_7, ν_8). Compared are all models shown in black, black symbol for median values, to the model with linear kinetics only (orange stars or green rectangles, white symbols for median values).

cillation probabilities are discussed in the main text, section 6.4.

For the negative feedback chain models, effects of combinations of reactions with saturating kinetics can be observed. If only all degradation reactions (Figure A.24 D) or only all conversion reactions (Figure A.24 C) exhibit saturating kinetics, the period and amplitude sensitivities are, if at all, only slightly increased. In contrast, if all seven reactions are allowed to saturate in the negative feedback chain model by using saturating kinetics, a very strong increase in period as well as amplitude sensitivities is observed (Figure A.24 A). Saturating or linear kinetics in reaction ν_2 seem not to affect the sensitivities considerably (compare Figure A.24 B to C).

For the positive feedback chain models, this investigation shows that saturating kinetics in neither all degradations only (Figure A.24 H) nor the inner-loop conversions only (Figure A.24 F) considerably affect the sensitivities. In contrast, saturating kinetics in ν_2 (and the inner-loop conversions) alters the sensitivities similarly to saturating kinetics in all reactions (Figure A.24 G, E). This is in accordance with the findings from section 6.3.5 that the saturation level of ν_2 yields the strongest impact on the sensitivities.

A.4.9. Matter flow and saturating kinetics

If employing saturating kinetics in the chain models C--- for the negative feedback and C2-- for the positive feedback, the according conversions are regulated productions. The analysis of the sensitivities of the models reveals that the overall period and amplitude sensitivities are slightly altered compared to the models C246 (Figure A.25 A, B). Still, the saturations of reactions ν_4 and ν_6 are most influential on the sensitivities in the negative feedback chain model (Figure A.25 D). Additionally, also for the positive feedback chain model, the saturation level of reaction ν_2 remains determining for both period and amplitude sensitivities (Figure A.25 E). However, in particular in the negative feedback chain model, altering the saturation of reactions ν_2, ν_3 and ν_7 does not exhibit significant changes in the period sensitivities in model C--- opposed to model C246 where the period sensitivities are significantly increased by the saturation of all single reactions (section 6.3.5). The only reactions whose saturation level affects the amplitude sensitivities significantly are ν_4, ν_6 and ν_7, i.e. reactions ν_2, ν_3 and ν_8 lost impact on the amplitude sensitivity whereas reaction ν_7 gained impact comparing models C246 and C---. For the positive feedback models, the impacts on the period sensitivities remain all significant for altering the saturation level from non-saturated to saturated. In contrast, for the amplitude sensitivities, the saturation levels of reactions ν_6 and ν_7 gain impact on the amplitude sensitivities, while that of reaction ν_8 loses its significant impact.

Altogether, altering the matter flow properties can affect the contribution of the saturation level of different reactions on the period and amplitude sensitivities, but the influence of the regulated productions are widely similar to the influences of the according conversions in the models with different matter flow properties.

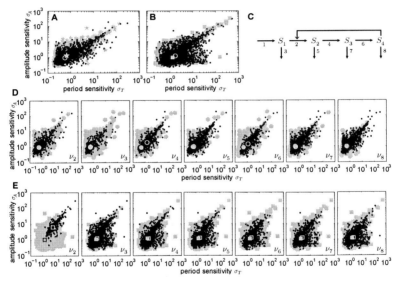

Figure A.25.: Effect of matter flow properties on saturation effects in the chain models with negative or positive feedback. A, B: Sensitivities for the model C--- with negative feedback and saturating kinetics (A) or C2-- with positive feedback and saturating kinetics (black dots, black circle for median values) compared to the according model C246 with saturating kinetics (light gray symbols, large symbol for median values). C: Structure of model C246. D, E: Sensitivities for reactions ν_2-ν_8 (one in each panel) being not saturated (light gray) or saturated (black) for model C--- with negative feedback and saturating kinetics (D) or model C2-- with positive feedback and saturating kinetics (E), large symbols represent median values.

A.5. Further applications

A.5.1. Sensitivity of synthetic oscillators

The sensitivities of the repressilator model as proposed in Elowitz and Leibler [2000] and of the metabolator proposed in Fung et al. [2005] are compared to the according models with saturating kinetics in all previously linear reactions (Figure A.26). Details of the models and the ODEs are provided in the Appendix, section A.7.7.

For both models, a significant increase in the amplitude sensitivities can be observed if (more) reactions with saturating kinetics are employed (median values change by 3-fold for the repressilator, Figure A.26 A, MWU p-value 0, and by 75% for the metabolator, Figure A.26 B, MWU p-value $3.3 \cdot 10^{-72}$, Tables A.20, A.21). For the period sensitivities, only the repressilator model displays significant alterations, the median period sensitivity increases by more than nine-fold (MWU p-value 0, change for the metabolator: $+10\%$, MWU p-value 0.30, Tables A.20, A.21).

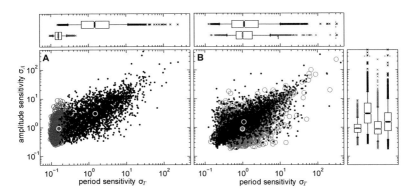

Figure A.26.: Impact of the kinetics on the sensitivities of synthetic oscillators. A: Sensitivities of the repressilator (Elowitz and Leibler [2000], brown circles, white framed circle for median values) compared to the sensitivities of the repressilator with saturating kinetics (black dots, black circle as median values). B: Sensitivities of the metabolator (Fung et al. [2005], violet circles, white framed circle for median values) compared to sensitivities for the metabolator with saturating kinetics (black dots, black circle as median values). The box-plots show the characteristics of the sensitivity distributions for period sensitivity (top) and amplitude sensitivity (right).

A.5.2. Sensitivity of an NF-κB model

Sensitivities of an NF-κB model derived by reduction of the model proposed in Kearns et al. [2006] are obtained from a sensitivity analysis as proposed in this thesis. Model details and the ODE system are provided in the Appendix, section A.7.8. Distributions characteristics are given in Table A.22.

In fact, both very high and very low period as well as amplitude sensitivities can be observed for this model. The fifth and 95th percentiles of the period sensitivity distribution lie at 0.15 and 1.95, respectively, but the values range from 0.091 to 30.4, which enables the occurrence of very low and very high period sensitivities for the model. Similarly, the fifth and 95th percentiles of the amplitude sensitivity distribution lie at 0.27 and 2.23, respectively, the values range from 0.17 to 25.5. Also for the amplitude, very low and very high sensitivities are obtained in the model.

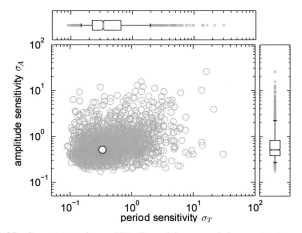

Figure A.27.: Sensitivities for an NF-κB model to reveal the sensitivity potential of the oscillations. Period and amplitude sensitivities of a model obtained by reducing the model proposed in Kearns et al. [2006] (green circles, white circle as median values).

A.6. Tables for sensitivity values

In the following tables, the detailed characteristics of the period sensitivity distributions (per.) and amplitude sensitivity distributions (ampl.) of the sensitivity analyses performed in this work are shown if not given together with the examination. The caption of each table delivers a short description of the captured models. Given are the identifier of the model (id.), the median sensitivity (μ), the 95% confidence interval of the median (95% CI), the interval formed by the first and third quartiles ($[Q_{25}, Q_{75}]$), the interval formed by the fifth and 95th percentiles ($[Q_5, Q_{95}]$), the length of the 90% data range (90%R), Spearman's rank correlation coefficient (R_S) for the correlation between period and amplitude sensitivity, the number of parameter sets being sampled in order to obtain 2500 parameter sets yielding regular oscillations (# sets).

The detailed test results of the MWU test (see section 2.4.5) indicating the significance of location differences between sensitivity distributions are indicated in separate tables. The p-value (MWU p), the numbers of parameter sets compared (n_X, n_Y), the calculated value of the U statistics (U) and the value for the normal distribution with according mean and standard deviation (z-val) are delivered.

A.6.1. Chain models with positive or negative feedback

Table A.8.: Sensitivity distribution characteristics of the chain models with mass action kinetics, matter flow and negative (neg fb.) or positive feedback (pos fb.).

id	sens.	μ	95% CI	$[Q_{25},Q_{75}]$	$[Q_5,Q_{95}]$	90%R	R_S	# sets
neg fb.	per.	0.19	0.19	[0.18,0.20]	[0.17,0.22]	0.05	0.30	$2.0 \cdot 10^7$
	ampl.	0.66	[0.65,0.68]	[0.50,0.93]	[0.35,1.84]	1.49		
pos fb.	per.	0.68	[0.67, 0.69]	[0.58,0.82]	[0.41,2.40]	1.99	0.42	$1.5 \cdot 10^5$
	ampl.	0.57	[0.56,0.57]	[0.51,0.65]	[0.43,4.91]	4.48		

Table A.9.: Results for the MWU test comparing the sensitivity distributions of the chain models with mass action kinetics, matter flow and negative (neg fb.) or positive feedback (pos fb.).

id_X		id_Y	sens.	MWU p	n_X	n_Y	U	z-val
neg fb.	$\Leftrightarrow$	pos fb.	per	0	2500	2500	1420	-61.20
pos fb.	$\Leftrightarrow$	neg fb.	ampl.	$2.5 \cdot 10^{-24}$	2500	2500	2609029	-10.11

A.6.2. Chain models with different matter flow properties

Sensitivity distribution characteristics

Table A.10.: Sensitivity distribution characteristics of the chain models with mass action kinetics, negative feedback and different matter flow properties. Matter flow in reactions ν_2, ν_4, ν_6 (identifier C246, original chain model with negative feedback); matter flow in ν_2, ν_4 (C24-); in ν_2, ν_6 (C2-6); in ν_2 (C2--); in ν_4, ν_6 (C-46); in ν_4 (C-4-); in ν_6 (C--6); in none of the three reactions (C---).

id	sens.	μ	95% CI	$[Q_{25},Q_{75}]$	$[Q_5,Q_{95}]$	90%R	R_S	# sets
C246	per.	0.189	[0.188,190]	[0.179,0.196]	[0.166,0.216]	0.05	0.30	$2.0 \cdot 10^7$
	ampl.	0.66	[0.65,0.68]	[0.50,0.93]	[0.35,1.84]	1.49		
C24-	per.	0.194	[0.194,0.195]	[0.187,0.199]	[0.175,0.209]	0.034	0.44	$9.9 \cdot 10^6$
	ampl.	0.71	[0.69,0.73]	[0.51,1.03]	[0.35,2.37]	2.02		
C2-6	per.	0.195	[0.194,0.195]	[0.190,0.198]	[0.176,0.202]	0.026	0.43	$9.5 \cdot 10^6$
	ampl.	0.64	[0.62,0.65]	[0.48,0.86]	[0.35,1.35]	1.00		
C2--	per.	0.200	[0.199,0.200]	[0.196,0.203]	[0.193,0.222]	0.029	0.79	$7.5 \cdot 10^6$
	ampl.	0.71	[0.69,0.73]	[0.51,1.03]	[0.37,2.21]	1.84		
C-46	per.	0.191	[0.190,0.192]	[0.182,0.197]	[0.167,0.202]	0.035	0.42	$9.9 \cdot 10^6$
	ampl.	0.67	[0.65,0.69]	[0.51,0.98]	[0.38,1.71]	1.33		
C-4-	per.	0.194	[0.194,0.195]	[0.189,0.199]	[0.175,0.203]	0.028	0.58	$5.0 \cdot 10^6$
	ampl.	0.70	[0.68,0.72]	[0.52,1.01]	[0.37,1.82]	1.45		
C--6	per.	0.194	[0.193,0.194]	[0.189,0.198]	[0.176,0.203]	0.027	0.48	$9.0 \cdot 10^6$
	ampl.	0.62	[0.61,0.64]	[0.48,0.89]	[0.36,1.64]	1.28		
C---	per.	0.198	0.198	[0.195,0.201]	[0.192,0.204]	0.012	0.94	$5.2 \cdot 10^6$
	ampl.	0.71	[0.69,0.73]	[0.52,1.04]	[0.36,1.79]	1.43		

Table A.11.: Sensitivity distribution characteristics of the chain models with mass action kinetics, positive feedback and different matter flow properties. Matter flow in reactions ν_2, ν_4, ν_6 (C246); in ν_2, ν_4 (C24-); in ν_2, ν_6 (C2-6); in ν_2 (C2--).

id	sens.	μ	95% CI	$[Q_{25},Q_{75}]$	$[Q_5,Q_{95}]$	90%R	R_S	# sets
C246	per.	0.68	[0.67,0.69]	[0.58,0.82]	[0.41,2.40]	1.99	0.42	$1.5 \cdot 10^5$
	ampl.	0.57	[0.56,0.57]	[0.51,0.65]	[0.43,4.91]	4.48		
C24-	per.	0.80	[0.80,0.81]	[0.67,0.98]	[0.48,2.37]	1.89	0.27	$3.7 \cdot 10^5$
	ampl.	0.63	[0.63,0.64]	[0.55,0.71]	[0.44,3.07]	2.63		
C2-6	per.	0.83	[0.83,0.84]	[0.74,1.03]	[0.52,2.74]	2.22	0.30	$7.4 \cdot 10^5$
	ampl.	0.59	[0.58,0.59]	[0.53,0.67]	[0.44,1.43]	1.01		
C2--	per.	0.93	[0.92,0.93]	[0.76,1.17]	[0.56,3.06]	2.50	0.14	$2.5 \cdot 10^5$
	ampl.	0.68	[0.67,0.69]	[0.56,0.79]	[0.45,5.65]	5.20		

Results of the MWU-test

Table A.12.: Results for the MWU test comparing the sensitivity distributions of the chain models with negative feedback and different matter flow properties. Matter flow in reactions ν_2, ν_4, ν_6 (C246); in ν_2, ν_4 (C24-); in ν_2, ν_6 (C2-6); in ν_2 (C2--); in ν_4, ν_6 (C-46); in ν_4 (C-4-); in ν_6 (C--6); in none of the three reactions (C---). The last six rows show the comparisons for the models differing only in the matter flow properties of reaction ν_2.

id_X		id_Y	sens.	MWU p	n_X	n_Y	U	z-val
C246	$\Leftrightarrow$	C24-	per.	$1.7 \cdot 10^{-57}$	2500	2500	2311578	-15.9382
			ampl.	$1.3 \cdot 10^{-5}$	2500	2500	2910577	-4.2014
C246	$\Leftrightarrow$	C2-6	per.	$1.2 \cdot 10^{-68}$	2500	2500	2233281	-17.4723
C2-6	$\Leftrightarrow$	C246	ampl.	$1.5 \cdot 10^{-5}$	2500	2500	2911883	-4.1758
C246	$\Leftrightarrow$	C2--	per.	0	2500	2500	1123067	-39.2258
			ampl.	$1.9 \cdot 10^{-7}$	2500	2500	2866066	-5.0735
C246	$\Leftrightarrow$	C-46	per.	$1.9 \cdot 10^{-4}$	2500	2500	2943734	-3.5517
			ampl.	0.0145	2500	2500	3013592	-2.1829
C246	$\Leftrightarrow$	C-4-	per.	$4.6 \cdot 10^{-59}$	2500	2500	2300107	-16.1629
			ampl.	$5.2 \cdot 10^{-5}$	2500	2500	2926937	-3.8808
C246	$\Leftrightarrow$	C--6	per.	$3.1 \cdot 10^{-52}$	2500	2500	2351096	-15.1638
C--6	$\Leftrightarrow$	C246	ampl.	$8.4 \cdot 10^{-5}$	2500	2500	2932961	-3.7628
C246	$\Leftrightarrow$	C---	per.	$1.5 \cdot 10^{-259}$	2500	2500	1369600	-34.3952
			ampl.	$3.6 \cdot 10^{-6}$	2500	2500	2895868	-4.4896
C-4-	$\Leftrightarrow$	C24-	per.	0.1943	2500	2500	3080990	-0.8623
			ampl.	0.3233	2500	2500	3101600	-0.4585
C--6	$\Leftrightarrow$	C2-6	per.	$5.7 \cdot 10^{-5}$	2500	2500	2928102	-3.8580
C2-6	$\Leftrightarrow$	C--6	ampl.	0.3793	2500	2500	3109318	-0.3073
C---	$\Leftrightarrow$	C2---	per.	$5.5 \cdot 10^{-40}$	2500	2500	2452177	-13.1833
			ampl.	0.2595	2500	2500	3092085	-0.6449

Table A.13.: Results for the MWU test comparing the sensitivity distributions of the chain models with positive feedback and different matter flow properties. Matter flow in ν_2, ν_4 and ν_6 (C246); in ν_2, ν_4 (C24-); in ν_2, ν_6 (C2-6); in ν_2 (C2--).

id_X		id_Y	sens.	MWU p	n_X	n_Y	U	z-val
C246	$\Leftrightarrow$	C24-	per.	$1.6 \cdot 10^{-63}$	2500	2500	2268439	-16.7834
			ampl.	$9.7 \cdot 10^{-45}$	2500	2500	2411276	-13.9847
C246	$\Leftrightarrow$	C2-6	per.	$8.8 \cdot 10^{-129}$	2500	2500	1894281	-24.1147
			ampl.	$6.9 \cdot 10^{-8}$	2500	2500	2856117	-5.2685
C246	$\Leftrightarrow$	C2--	per.	$3.3 \cdot 10^{-180}$	2500	2500	1665321	-28.6009
			ampl.	$7.5 \cdot 10^{-89}$	2500	2500	2106821	-19.9502

A.6.3. Chain models with saturating kinetics

Table A.14.: Sensitivity distribution characteristics of the chain models with mass action kinetics (ma) or Michaelis-Menten kinetics (MM), matter flow and negative (neg fb.) or positive feedback (pos fb.). The data for the models with mass action kinetics can be also found in Table A.8.

id	sens.	μ	95% CI	$[Q_{25},Q_{75}]$	$[Q_5,Q_{95}]$	90%R	R_S	# sets
neg fb.	per.	0.19	[0.19,0.19]	[0.18,0.20]	[0.17,0.22]	0.05	0.30	$2.0 \cdot 10^7$
ma	ampl.	0.66	[0.65,0.68]	[0.50,0.93]	[0.35,1.84]	1.49		
neg fb.	per.	0.66	[0.63,0.71]	[0.35,1.64]	[0.23,6.55]	6.32	0.76	$1.4 \cdot 10^6$
MM	ampl.	1.34	[1.26,1.41]	[0.71,3.01]	[0.43,8.97]	8.54		
pos fb.	per.	1.33	[1.26, 1.40]	[0.70,3.26]	[0.42,13.34]	12.92	0.64	$1.7 \cdot 10^6$
MM	ampl.	0.98	[0.94,1.03]	[0.60,2.38]	[0.43,11.71]	11.28		
pos fb.	per.	0.68	[0.67, 0.69]	[0.58,0.82]	[0.41,2.40]	1.99	0.42	$1.5 \cdot 10^5$
ma	ampl.	0.57	[0.56,0.57]	[0.51,0.65]	[0.43,4.91]	4.48		

Table A.15.: Results for the MWU test comparing the sensitivity distributions of the chain models with matter flow and mass action kinetics (ma) or Michaelis-Menten kinetics (MM), negative feedback (neg fb.) or positive feedback (pos fb.).

id$_X$		id$_Y$	sens.	MWU p	n_X	n_Y	U	z-val
neg fb. ma	$\Leftrightarrow$	neg fb. MM	per.	0	2500	2500	39848	-60.45
			ampl.	$1.6 \cdot 10^{-207}$	2500	2500	1557243	-30.72
pos fb. ma	$\Leftrightarrow$	pos fb. MM	per.	$3.8 \cdot 10^{-173}$	2500	2500	1694628	-28.03
			ampl.	$1.2 \cdot 10^{-175}$	2500	2500	1684145	-28.23
neg fb. MM	$\Leftrightarrow$	pos fb. MM	per.	$7.2 \cdot 10^{-117}$	2500	2500	1953660	-22.95
pos fb. MM	$\Leftrightarrow$	neg fb. MM	ampl.	$6.9 \cdot 10^{-14}$	2500	2500	2747435	-7.40

A.6.4. Models of Tsai et al. and alterations

Table A.16.: Sensitivity distribution characteristics of the models of Tsai et al. and alterations. Given are the values for the negative feedback only model (tsai neg), for the model with a positive autoregulatory feedback (tsai pos), for the negative feedback model with Michaelis-Menten kinetics for the degradation reactions (tsai neg MM).

id	sens.	μ	95% CI	$[Q_{25},Q_{75}]$	$[Q_5,Q_{95}]$	90%R	R_S	# sets
tsai	per.	0.23	[0.23,0.23]	[0.21,0.26]	[0.21,0.32]	0.11	0.80	$9.7 \cdot 10^5$
neg	ampl.	0.49	[0.46,0.51]	[0.30,0.89]	[0.21,2.30]	2.09		
tsai	per.	0.28	[0.27,0.29]	[0.21,0.47]	[0.19,1.75]	1.56	0.24	$8.0 \cdot 10^5$
pos	ampl.	0.42	[0.41,0.44]	[0.24,0.90]	[0.01,3.30]	3.29		
tsai	per.	0.29	[0.29, 0.30]	[0.24,0.43]	[0.20,2.23]	2.03	0.50	$4.0 \cdot 10^5$
neg MM	ampl.	0.42	[0.41,0.44]	[0.30,0.80]	[0.21,2.95]	2.74		

Table A.17.: Results for the MWU test comparing the sensitivity distributions of the model of Tsai et al. with negative feedback only (tsai neg) to the model with additional positive feedback (tsai pos) and to the model with negative feedback and Michaelis-Menten kinetics for the degradation reactions (tsai neg MM), and the comparison of the sensitivities of the latter two models.

id_X		id_Y	sens.	MWU p	n_X	n_Y	U	z-val
tsai neg	$\Leftrightarrow$	tsai pos	per.	$2.2 \cdot 10^{-64}$	2500	2500	2262396	-16.90
tsai pos	$\Leftrightarrow$	tsai neg	ampl.	$6.9 \cdot 10^{-10}$	2500	2500	2815844	-6.06
tsai neg	$\Leftrightarrow$	tsai neg MM	per.	$2.8 \cdot 10^{-183}$	2500	2500	1652812	-28.84
tsai neg MM	$\Leftrightarrow$	tsai neg	ampl.	0.0064	2500	2500	2997831	-2.49
tsai pos	$\Leftrightarrow$	tsai neg MM	per.	$4.3 \cdot 10^{-7}$	2500	2500	2873783	-4.92
tsai pos	$\Leftrightarrow$	tsai neg MM	ampl.	$2.7 \cdot 10^{-5}$	2500	2500	2919145	-4.03

A.6.5. Circadian and calcium oscillations models

Table A.18.: Sensitivity distribution characteristics of the circadian rhythm models in Becker-Weimann et al. [2004] (id "circ M"), Goldbeter [1995] (id "circ D"), Locke et al. [2005b] (id "circ A"); of the calcium oscillation models in Goldbeter et al. [1990] (id "ca phen'), Sneyd et al. [2004] (id "ca open") and De Young and Keizer [1992] (id "ca closed"). Additionally to the usual values, the sensitivities obtained for the reference parameter set published together with the model are indicated (σ^*).

id	sens	σ^*	μ	95% CI	$[Q_{25}, Q_{75}]$	$[Q_5, Q_{95}]$	90%R	R_S	# sets
circ M	per.	0.11	0.14	[0.14,0.14]	[0.13,0.15]	[0.11,0.16]	0.05	0.52	$6.1 \cdot 10^5$
	ampl.	0.62	0.47	[0.46,0.48]	[0.35,0.68]	[0.24,1.37]	1.13		
circ D	per.	0.27	0.49	[0.45,0.53]	[0.26,1.49]	[0.20,8.03]	7.83	0.81	$1.7 \cdot 10^5$
	ampl.	1.11	0.97	[0.91,1.03]	[0.52,2.45]	[0.35,9.64]	9.29		
circ A	per.	0.22	0.44	[0.42,0.46]	[0.24,1.02]	[0.15,4.23]	4.08	0.65	$3.5 \cdot 10^5$
	ampl.	2.80	0.96	[0.91,1.03]	[0.46,2.44]	[0.25,8.67]	8.42		
ca phen	per.	1.64	1.16	[1.14,1.17]	[0.85,2.22]	[0.53,6.45]	5.92	0.82	$7.1 \cdot 10^4$
	ampl.	1.59	1.01	[1.00,1.03]	[0.74,1.94]	[0.46,6.54]	6.08		
ca open	per.	0.55	0.46	[0.46,0.47]	[0.37,0.63]	[0.29,2.62]	2.33	0.61	$5.9 \cdot 10^4$
	ampl.	0.22	0.36	[0.35,0.36]	[0.26,0.48]	[0.17,1.12]	0.95		
ca closed	per.	0.30	0.55	[0.53,0.57]	[0.36,0.87]	[0.19,2.25]	2.04	0.16	$1.5 \cdot 10^5$
	ampl.	1.33	0.36	[0.35,0.37]	[0.21,0.57]	[0.12,1.79]	1.67		

Table A.19.: Results for the MWU test comparing the period sensitivity distributions between the circadian oscillations models (circ), between the calcium oscillations models (ca), and between calcium and circadian oscillations models (ca/circ), as well as the according amplitude sensitivity distributions. Circadian models: Becker-Weimann et al. [2004] (id "circ M"), Goldbeter [1995] (id "circ D"), Locke et al. [2005b] (id "circ A"). Calcium models: Goldbeter et al. [1990] (id "ca phen"), Sneyd et al. [2004] (id "ca open") and De Young and Keizer [1992] (id "ca closed").

	id_X		id_Y	sens.	MWU p	n_X	n_Y	U	z-val
	circ M	⇔	circ D	per.	0	2500	2500	6642	-61.1
circ	circ M	⇔	circ A	per.	0	2500	2500	238492	-56.6
	circ A	⇔	circ D	per.	$7.5 \cdot 10^{-11}$	2500	2500	2798087	-6.4
	ca open	⇔	ca phen	per.	0	2500	2500	850223	-44.5
ca	ca closed	⇔	ca phen	per.	0	2500	2500	1189860	-37.9
	ca open	⇔	ca closed	per	$5.2 \cdot 10^{-10}$	2500	2500	2813457	-6.1
	circ M	⇔	ca phen	per.	0	2500	2500	0	-61.23
	circ M	⇔	ca open	per.	0	2500	2500	101	-61.2
	circ M	⇔	ca closed	per.	0	2500	2500	74751	-59.8
ca/	circ D	⇔	ca phen	per.	$6.7 \cdot 10^{-155}$	2500	2500	1773167	-26.5
circ	circ D	⇔	ca open	per.	0.13	2500	2500	3067850	-1.1
	circ D	⇔	ca closed	per.	0.20	2500	2500	3081382	-0.85
	circ A	⇔	ca phen	per.	$5.2 \cdot 10^{-251}$	2500	2500	1399002	-33.8
	circ A	⇔	ca open	per.	$1.0 \cdot 10^{-8}$	2500	2500	2838595	-5.6
	circ A	⇔	ca closed	per.	$3.2 \cdot 10^{-12}$	2500	2500	2774426	-6.9
	circ M	⇔	circ D	ampl.	$3.9 \cdot 10^{-238}$	2500	2500	1444295	-32.9
circ	circ M	⇔	circ A	ampl.	$4.6 \cdot 10^{-159}$	2500	2500	1754824	-26.8
	circ A	⇔	circ D	ampl.	$8.8 \cdot 10^{-4}$	2500	2500	2965299	-3.1
	ca open	⇔	ca phen	ampl.	0	2500	2500	522624	-51.0
ca	ca closed	⇔	ca phen	ampl.	0	2500	2500	824589	-45.1
	ca closed	⇔	ca open	ampl	0.089	2500	2500	3056252	-1.3
	circ M	⇔	ca phen	ampl.	0	2500	2500	978813	-42.1
	ca open	⇔	circ M	ampl.	$3.4 \cdot 10^{-96}$	2500	2500	2064547	-20.8
	ca closed	⇔	circ M	ampl.	$2.1 \cdot 10^{-68}$	2500	2500	2234986	-17.4
ca/	circ D	⇔	ca phen	ampl.	$2.8 \cdot 10^{-5}$	2500	2500	2919224	-4.0
circ	ca open	⇔	circ D	ampl.	0	2500	2500	855800	-44.4
	ca closed	⇔	circ D	ampl.	0	2500	2500	1099431	-39.7
	circ A	⇔	ca phen	ampl.	$7.6 \cdot 10^{-9}$	2500	2500	2836190	-5.7
	ca open	⇔	circ A	ampl.	0	2500	2500	1193590	-37.8
	ca closed	⇔	circ A	ampl.	$1.3 \cdot 10^{-270}$	2500	2500	1332216	-35.1

A.6.6. Synthetic oscillators

Table A.20.: Sensitivity distributions characteristics of the repressilator (Elowitz and Leibler [2000], id "repr"), the repressilator with saturating kinetics (id "repr MM"), the metabolator (Fung et al. [2005], id "metab") and the metabolator with saturating kinetics (id "metab MM"). Values correspond to Figure A.26.

id	sens	μ	95% CI	$[Q_{25},Q_{75}]$	$[Q_5,Q_{95}]$	90%R	R_S	# sets
repr	per.	0.16	[0.16,0.16]	[0.13,0.20]	[0.11,0.27]	0.16	0.45	$2.4 \cdot 10^7$
	ampl.	0.92	[0.90,0.94]	[0.67,1.29]	[0.45,2.27]	1.82		
repr MM	per.	1.48	[1.38,1.57]	[0.65,3.58]	[0.29,11.41]	11.12	0.65	$5.6 \cdot 10^6$
	ampl.	3.06	[2.86,3.26]	[1.41,7.09]	[0.55,17.97]	17.42		
metab	per.	0.98	[0.95,1.02]	[0.66,1.78]	[0.38,5.42]	5.04	0.40	$1.4 \cdot 10^6$
	ampl.	0.88	[0.85,0.91]	[0.59,1.58]	[0.33,5.16]	4.83		
metab MM	per.	1.08	[1.02,1.16]	[0.51,2.57]	[0.24,10.11]	9.87	0.65	$2.6 \cdot 10^6$
	ampl.	1.55	[1.49,163]	[0.76,3.21]	[0.37,10.29]	9.92		

Table A.21.: Results for the MWU test for comparing the sensitivities of the repressilator model (Elowitz and Leibler [2000], id "repr") or the metabolator model (Fung et al. [2005], id "metab") with the respective model with saturating kinetics (id "repr MM" or "metab MM").

id_X		id_Y	sens.	MWU p	n_X	n_Y	U	z-val
repr	$\Leftrightarrow$	repr MM	per.	0	2500	2500	44392	-60.4
			ampl.	0	2500	2500	1067165	-40.3
metab	$\Leftrightarrow$	metab MM	per.	0.30	2500	2500	3098075	-0.5
			ampl.	$3.3 \cdot 10^{-72}$	2500	2500	2209779	-17.9

A.6.7. NF-κB oscillations model

Table A.22.: Sensitivity distribution characteristics of the NF-κB model derived by reduction from Kearns et al. [2006] as displayed in Figure A.27.

id	sens	μ	95% CI	$[Q_{25},Q_{75}]$	$[Q_5,Q_{95}]$	90%R	R_S	# sets
NF-κB	per.	0.34	[0.32,0.35]	[0.22,0.65]	[0.15,1.95]	1.80	0.36	$1.4 \cdot 10^5$
	ampl.	0.52	[0.50,0.53]	[0.38,0.83]	[0.27,2.23]	1.96		

A.7. Model equations

A.7.1. Circadian and calcium oscillations models

Becker-Weimann et al. [2004]: Mammalian circadian rhythm

This ODE system models the circadian rhythm of mammals as given in Becker-Weimann et al. [2004]. Therein, the heterodimer formed by PER2 and CRY (but not the separate entities) and the protein BMAL1 are considered. The nuclear PER2/CRY complex inhibits transcription of the *per2* and/or *cry* genes and positively regulates transcription of the *bmal1* gene, BMAL1 protein is produced in the cytosol, is reversibly transported to the nucleus and gets reversibly activated. The activated BMAL1 protein enhances *per2/cry* mRNA production, from the mRNA cytosolic PER2/CRY complex is produced and can reversibly enter the nucleus. There are 7 different species, connected by 17 reactions (hence, there are 17 rate coefficients) having 4 nl-parameters and 3 cooperativity parameters.

$$\frac{dS_1}{dt} = k_1 \cdot (S_7 + kn_1)/(kn_2 \cdot (1 + (S_3/kn_3)^{n_1}) + S_7 + kn_1) - k_2 \cdot S_1$$

$$\frac{dS_2}{dt} = k_3 \cdot S_1^{n_2} - k_4 \cdot S_2 - k_5 \cdot S_2 + k_6 \cdot S_3$$

$$\frac{dS_3}{dt} = k_5 \cdot S_2 - k_6 \cdot S_3 - k_7 \cdot S_3$$

$$\frac{dS_4}{dt} = k_8 \cdot S_3^{n_3}/(kn_4^{n_3} + S_3^{n_3}) - k_9 \cdot S_4$$

$$\frac{dS_5}{dt} = k_{10} \cdot S_4 - k_{11} \cdot S_5 - k_{12} \cdot S_5 + k_{13} \cdot S_6$$

$$\frac{dS_6}{dt} = k_{12} \cdot S_5 - k_{13} \cdot S_6 - k_{14} \cdot S_6 - k_{15} \cdot S_6 + k_{16} \cdot S_7$$

$$\frac{dS_7}{dt} = k_{15} \cdot S_6 - k_{16} \cdot S_7 - k_{17} \cdot S_7$$

In the original publication by Becker-Weimann et al. [2004], the notation and parameter values given in the following table are used.

S_1	$y1$		k_5	k_{2t}	$0.24\ h^{-1}$	k_{16}	k_{7a}	$0.003\ h^{-1}$
S_2	$y2$		k_6	k_{3t}	$0.02\ h^{-1}$	k_{17}	k_{7d}	$0.09\ h^{-1}$

S_3	$y3$		k_7	k_{3d}	$0.12\ h^{-1}$	kn_1	c	$0.01\ nM$
S_4	$y4$		k_8	ν_{4b}	$3.6\ nMh^{-1}$	kn_2	k_{1b}	$1\ nM$
S_5	$y5$		k_9	k_{4d}	$0.75\ h^{-1}$	kn_3	k_{1i}	$0.56\ nM$
S_6	$y6$		k_{10}	k_{5b}	$0.24\ h^{-1}$	kn_4	k_{4b}	$2.16\ nM$
S_7	$y7$		k_{11}	k_{5d}	$0.06\ h^{-1}$	n_1	p	8
k_1	ν_{1b}	$9\ nMh^{-1}$	k_{12}	k_{5t}	$0.45\ h^{-1}$	n_2	q	2
k_2	k_{1d}	$0.12\ h^{-1}$	k_{13}	k_{6t}	$0.06\ h^{-1}$	n_3	r	3
k_3	k_{2b}	$0.3\ nM^{-1}h^{-1}$	k_{14}	k_{6d}	$0.12\ h^{-1}$			
k_4	k_{2d}	$0.05\ h^{-1}$	k_{15}	k_{6a}	$0.09\ h^{-1}$			

Goldbeter [1995]: *Drosophila* circadian rhythm

This ODE system models the circadian rhythm of the fruit fly (*Drosophila*) as given in Goldbeter [1995]. It is a minimal model considering the *per* mRNA, the PER protein, two reversible phosphorylation steps of it and a reversible transport to the nucleus where it feeds back negatively on the transcription of the *per* mRNA. The model contains 5 different species, 10 rate coefficients, 7 nl-parameters and 1 cooperativity parameter. Reaction indices can be derived from the index of their according rate coefficient.

$$\frac{dS_1}{dt} = k_1 \cdot kn_1^{n_1}/(kn_1^{n_1} + S_5^{n_1}) - k_2 \cdot S_1/(kn_2 + S_1)$$

$$\frac{dS_2}{dt} = k_3 \cdot S_1 - k_4 \cdot S_2/(kn_3 + S_2) + k_5 \cdot S_3/(kn_4 + S_3)$$

$$\frac{dS_3}{dt} = k_4 \cdot S_2/(kn_3 + S_2) - k_5 \cdot S_3/(kn_4 + S_3) - k_6 \cdot S_3/(kn_5 + S_3)$$
$$+ k_7 \cdot S_4/(kn_6 + S_4)$$

$$\frac{dS_4}{dt} = k_6 \cdot S_3/(kn_5 + S_3) - k_7 \cdot S_4/(kn_6 + S_4) - k_8 \cdot S_4 + k_9 \cdot S_5$$
$$- k_{10} \cdot S_4/(kn_7 + S_4)$$

$$\frac{dS_5}{dt} = k_8 \cdot S_4 - k_9 \cdot S_5$$

The notation, initial conditions and values for the parameters used in the original publication (Goldbeter [1995]) are given in the table below.

S_1	M	$0.1~\mu M$	k_4	V_1	$3.2~\mu M h^{-1}$	kn_2	K_m	$0.5~\mu M$
S_2	P_0	$0.25~\mu M$	k_5	V_2	$1.58~\mu M h^{-1}$	kn_3	K_1	$2~\mu M$
S_3	P_1	$0.25~\mu M$	k_6	V_3	$5~\mu M h^{-1}$	kn_4	K_2	$2~\mu M$
S_4	P_2	$0.25~\mu M$	k_7	V_4	$2.5~\mu M h^{-1}$	kn_5	K_3	$2~\mu M$
S_5	P_N	$0.25~\mu M$	k_8	k_1	$1.9~h^{-1}$	kn_6	K_4	$2~\mu M$
k_1	ν_s	$0.76~\mu M h^{-1}$	k_9	k_2	$1.3~h^{-1}$	kn_7	K_d	$0.2~\mu M$
k_2	ν_m	$0.65~\mu M h^{-1}$	k_{10}	ν_d	$0.95~\mu M h^{-1}$	n_1	n	4
k_3	k_s	$0.38~h^{-1}$	kn_1	K_I	$1~\mu M$			

Locke et al. [2005]: *Arabidopsis* circadian rhythm

This model encodes the circadian rhythm of arabidopsis (mouse-ear cress, *Arabidopsis thaliana*) and is published as model2, the interlocked feedback model, in Locke et al. [2005b] which is based on a model in Locke et al. [2005a] and is an extension of model1 in Locke et al. [2005b]. Therein, the light enters as explicit parameter. In this thesis, the model is investigated for constant light or darkness condition. In the case of constant light, species 13 ($c_P^{(n)}$) from the original model is only degraded, it gets eventually zero for all possible parameter sets, and thus does not enter into any of the other equations after a certain integration time. In the case of constant darkness, the concentration of species 13 is completely decoupled from the rest of the ODE system. Therefore, the model is restricted to the 12 other species. The light term (being constant in time) is not explicitly included.

Modeled are the proteins LHY (late elongated hypocotyl), TOC1 (timing of cab 1) and two unknown proteins X and Y with their mRNA, their cytosolic and nuclear concentrations each. For all four proteins, the according mRNA is produced, cytosolic protein is translated from it which is then reversibly transported to the nucleus. The mRNA, nuclear and cytosolic protein concentrations are subject to degradations. These three species each for LHY, TOC1 and X are connected in a negative feedback loop in which nuclear LHY protein acts negatively on TOC1 mRNA, nuclear TOC1 protein or nuclear X protein acts positively on X mRNA or LHY mRNA, respectively. The Y mRNA production is inhibited by both nuclear LHY protein as well as nuclear TOC1 protein. Nuclear Y protein acts positively on TOC1 mRNA production. The model consists of 12 species, 28 reactions, 18 nl-parameters and 6 cooperativity parameters.

$$\frac{\mathrm{d}S_1}{\mathrm{d}t} = k_1 \cdot S_9^{n_1}/(kn_1^{n_1} + S_9^{n_1}) - k_2 \cdot S_1/(S_1 + kn_2)$$

$$\frac{\mathrm{d}S_2}{\mathrm{d}t} = k_3 \cdot S_1 - k_4 \cdot S_2 + k_5 \cdot S_3 - k_6 \cdot S_2/(S_2 + kn_3)$$

$$\frac{\mathrm{d}S_3}{\mathrm{d}t} = k_4 \cdot S_2 - k_5 \cdot S_3 - k_7 \cdot S_3/(S_3 + kn_4)$$

$$\frac{\mathrm{d}S_4}{\mathrm{d}t} = k_8 \cdot S_{12}^{n_2}/(kn_5^{n_2} + S_{12}^{n_2}) \cdot kn_6^{n_3}/(kn_6^{n_3} + S_3^{n_3}) - k_9 \cdot S_4/(S_4 + kn_7)$$

$$\frac{\mathrm{d}S_5}{\mathrm{d}t} = k_{10} \cdot S_4 - k_{11} \cdot S_5 + k_{12} \cdot S_6 - k_{13} \cdot S_5/(S_5 + kn_8)$$

$$\frac{\mathrm{d}S_6}{\mathrm{d}t} = k_{11} \cdot S_5 - k_{12} \cdot S_6 - k_{14} \cdot S_6/(S_6 + kn_9)$$

$$\frac{\mathrm{d}S_7}{\mathrm{d}t} = k_{15} \cdot S_6^{n_4}/(kn_{10}^{n_4} + S_6^{n_4}) - k_{16} \cdot S_7/(S_7 + kn_{11})$$

$$\frac{\mathrm{d}S_8}{\mathrm{d}t} = k_{17} \cdot S_7 - k_{18} \cdot S_8 + k_{19} \cdot S_9 - k_{20} \cdot S_8/(S_8 + kn_{12})$$

$$\frac{\mathrm{d}S_9}{\mathrm{d}t} = k_{18} \cdot S_8 - k_{19} \cdot S_9 - k_{21} \cdot S_9/(S_9 + kn_{13})$$

$$\frac{\mathrm{d}S_{10}}{\mathrm{d}t} = k_{22} \cdot kn_{14}^{n_5}/(kn_{14}^{n_5} + S_6^{n_5}) \cdot kn_{15}^{n_6}/(kn_{15}^{n_6} + S_3^{n_6}) - k_{23} \cdot S_{10}/(S_{10} + kn_{16})$$

$$\frac{\mathrm{d}S_{11}}{\mathrm{d}t} = k_{24} \cdot S_{10} - k_{25} \cdot S_{11} + k_{26} \cdot S_{12} - k_{27} \cdot S_{11}/(S_{11} + kn_{17})$$

$$\frac{\mathrm{d}S_{12}}{\mathrm{d}t} = k_{25} \cdot S_{11} - k_{26} \cdot S_{12} - k_{28} \cdot S_{12}/(S_{12} + kn_{18})$$

Parameter pairs which are distinguished only by forming a product with the light term Θ or not are coded as one single parameter (see table below). In the original publication (Locke et al. [2005b]), the following notation and parameter values are used (units being nMh^{-1} or h^{-1} for rate coefficients, nM for nl-parameters, and no unit for cooperativity parameters). The initial conditions are not supplied. In the table, the values for the species give the calculated concentration (in nM) at the unstable steady state for $\Theta = const = 0$ (constant dark).

S_1	$c_L^{(m)}$	0.1960	k_{10}	p_2	4.3240	kn_3	k_2	1.5644
S_2	$c_L^{(c)}$	0.0056	k_{11}	r_3	0.3166	kn_4	k_3	1.2765

S_3	$c_L^{(n)}$	0.0317	k_{12}	r_4	2.1509	kn_5	g_2	0.0368
S_4	$c_T^{(m)}$	1.1746	k_{13}	$(1-\Theta(t))m_5+m_6$	0.0013, 3.1741	kn_6	g_3	0.2658
S_5	$c_T^{(c)}$	12.1230	k_{14}	$(1-\Theta(t))m_7+m_8$	0.0492, 4.4024	kn_7	k_4	2.5734
S_6	$c_T^{(n)}$	0.6269	k_{15}	n_3	0.2431	kn_8	k_5	2.7454
S_7	$c_X^{(m)}$	0.0886	k_{16}	m_9	10.1132	kn_9	k_6	0.4033
S_8	$c_X^{(c)}$	1.0209	k_{17}	p_3	2.1470	kn_{10}	g_4	0.5388
S_9	$c_X^{(n)}$	0.3025	k_{18}	r_5	1.0352	kn_{11}	k_7	6.5585
S_{10}	$c_Y^{(m)}$	0.0416	k_{19}	r_6	3.3017	kn_{12}	k_8	0.6632
S_{11}	$c_Y^{(c)}$	0.0071	k_{20}	m_{10}	0.2179	kn_{13}	k_9	17.1111
S_{12}	$c_Y^{(n)}$	0.0295	k_{21}	m_{11}	3.3442	kn_{14}	g_5	1.1780
k_1	n_1	5.1694	k_{22}	$\Theta(t)n_4+n_5$	0.0857, 0.1649	kn_{15}	g_6	0.0645
k_2	m_1	1.5283	k_{23}	m_{12}	4.2970	kn_{16}	k_{10}	1.7303
k_3	p_1	0.8295	k_{24}	p_4	0.2485	kn_{17}	k_{11}	1.8258
k_4	r_1	16.8363	k_{25}	r_7	2.2123	kn_{18}	k_{12}	1.8066
k_5	r_2	0.1687	k_{26}	r_8	0.2002	n_1	a	3.3064
k_6	m_2	20.4400	k_{27}	m_{13}	0.1347	n_2	b	1.0258
k_7	m_3	3.6888	k_{28}	m_{14}	0.6114	n_3	c	1.0258
k_8	n_2	3.0087	kn_1	g_1	0.8767	n_4	d	1.4422
k_9	m_4	3.8231	kn_2	k_1	1.8170	n_5	e	3.6064
						n_6	f	1.0237

Goldbeter et al. [1990]: Phenomenological calcium oscillations

The model published in Goldbeter et al. [1990] is a phenomenological model of cellular calcium oscillations. Only the cytosolic calcium concentration and the calcium concentration in the ER are considered as species. Cytosolic calcium is constantly produced (by two constant rates, one of which depends on a saturation function β by which the effect of IP$_3$ is included) and linearly degraded. One process represents pumping of calcium from the cytosol to the ER (depending in a Hill-function from cytosolic calcium), and two processes the release of calcium from the ER to the cytosol: one linear basal rate ("leaky transport") and one rate depending on both cytosolic ("activation" part) and endoplasmic ("release" part) calcium in a Hill-type manner. The model yields 2 species, 6 rate coefficients, 3 nl-parameters and 3 cooperativity parameters.

$$\frac{\mathrm{d}S_1}{\mathrm{d}t} = k_1 + k_2 - k_3 \cdot \frac{S_1^{n_1}}{kn_1^{n_1} + S_1^{n_1}} + k_4 \cdot \frac{S_2^{n_2} \cdot S_1^{n_3}}{(kn_2^{n_2} + S_2^{n_2}) \cdot (kn_3^{n_3} + S_1^{n_3})}$$
$$\qquad + k_5 \cdot S_2 - k_6 \cdot S_1$$
$$\frac{\mathrm{d}S_2}{\mathrm{d}t} = k_3 \cdot \frac{S_1^{n_1}}{kn_1^{n_1} + S_1^{n_1}} - k_4 \cdot \frac{S_2^{n_2} \cdot S_1^{n_3}}{(kn_2^{n_2} + S_2^{n_2}) \cdot (kn_3^{n_3} + S_1^{n_3})} - k_5 \cdot S_2$$

The notation and values employed in Goldbeter et al. [1990] are given in the following table. The saturation β is assumed to take a value of 0.4 for estimation of the sensitivity at the reference parameter set. Species concentrations denote steady state levels (of the unstable steady state) for the given reference parameter set.

S_1	Z	$0.3920\ \mu M$	k_4	V_{M3}	$500\ \mu M s^{-1}$	kn_3	K_A	$0.9\ \mu M$
S_2	Y	$1.6456\ \mu M$	k_5	k_f	$1\ s^{-1}$	n_1	n	2
k_1	ν_0	$1\ \mu M s^{-1}$	k_6	k	$10\ s^{-1}$	n_2	m	2
k_2	$\nu_1 \cdot \beta$	$7.3\ \mu M s^{-1}$, $(0.3, 0.77)$	kn_1	K_2	$1\ \mu M$	n_3	p	4
k_3	V_{M2}	$65\ \mu M s^{-1}$	kn_2	K_R	$2\ \mu M$			

Sneyd et al. [2004]: Open cell calcium oscillations

This calcium oscillations model is based on a model proposed in Sneyd and Dufour [2002]. It considers the cytosolic calcium concentration as well as the total amount of moles in the cell divided by the cell volume. It is an open cell model which means that it allows for influx of calcium into the cell (at two constant rates, one of which depending on the constant concentration of IP$_3$) and efflux from the cell (modeled by a Hill-kinetics term depending on the cytosolic calcium concentration). The other five variables display the probabilities of different states of the IP$_3$-receptor, their sum is below or equal to one, the missing amount to one is the probability of a sixth (dependent) state which is therefore not included in the model. The receptor allows for calcium flux to the cytosol (from the ER or comparable compartments) in only two states, S_4 and S_5, this calcium flux assumed to be proportional to $(0.1S_4 + 0.9S_5)^n$ which is not varied in the parameter sampling process. The model contains 7 variables, 24 rate coefficients, 7 nl-parameters and 2 cooperativity parameters.

$$\frac{dS_1}{dt} = k_1 \cdot (0.1 \cdot S_4 + 0.9 \cdot S_5)^{n_1} \cdot (\frac{S_2}{kn_7} - (1 + \frac{1}{kn_7}) \cdot S_1) + k_2 \cdot (\frac{S_2}{kn_7} - (1 + \frac{1}{kn_7}) \cdot S_1)$$
$$- k_3 \cdot \frac{S_1}{kn_1 + S_1} \cdot \frac{kn_7}{S_2 - S_1} + k_4 + k_5 \cdot kn_3 - k_6 \cdot \frac{S_1^{n_2}}{kn_2^{n_2} + S_1^{n_2}}$$

$$\frac{dS_2}{dt} = k_4 + k_5 \cdot kn_3 - k_6 \cdot \frac{S_1^{n_3}}{kn_2^{n_3} + S_1^{n_3}}$$

$$\frac{dS_3}{dt} = k_7 \cdot \frac{kn_6}{kn_6 + S_1} \cdot S_4 + k_8 \cdot \frac{kn_6 \cdot S_1}{kn_6 + S_1} \cdot S_4 - k_9 \cdot \frac{kn_4 \cdot kn_5 \cdot kn_3 \cdot S_3}{kn_4 \cdot kn_5 + S_1 \cdot (kn_4 + kn_5)}$$
$$- k_{10} \cdot \frac{kn_4 \cdot S_1 \cdot S_3 \cdot kn_3}{kn_4 \cdot kn_5 + S_1 \cdot (kn_4 + kn_5)} + k_{11} \cdot S_6 + k_{12} \cdot S_6$$
$$- k_{13} \cdot \frac{kn_4 \cdot kn_5 \cdot S_1 \cdot S_3}{kn_4 \cdot kn_5 + S_1 \cdot (kn_4 + kn_5)} - k_{14} \cdot \frac{kn_5 \cdot S_1 \cdot S_3}{kn_4 \cdot kn_5 + S_1 \cdot (kn_4 + kn_5)}$$

$$\frac{dS_4}{dt} = k_9 \cdot \frac{kn_4 \cdot kn_5 \cdot kn_3 \cdot S_3}{kn_4 \cdot kn_5 + S_1 \cdot (kn_4 + kn_5)} + k_{10} \cdot \frac{kn_4 \cdot S_1 \cdot S_3 \cdot kn_3}{kn_4 \cdot kn_5 + S_1 \cdot (kn_4 + kn_5)}$$
$$- k_7 \cdot \frac{kn_6}{kn_6 + S_1} \cdot S_4 - k_8 \cdot \frac{kn_6 \cdot S_1}{kn_6 + S_1} \cdot S_4 - k_{15} \cdot \frac{kn_6 \cdot S_1 \cdot S_4}{kn_6 + S_1} - k_{16} \cdot \frac{S_1 \cdot S_4}{kn_6 + S_1}$$
$$- k_{17} \cdot \frac{kn_6 \cdot S_4}{kn_6 + S_1} + k_{18} \cdot \frac{kn_4 \cdot S_5}{kn_4 + S_1} + k_{19} \cdot \frac{kn_4 \cdot S_5}{kn_4 + S_1}$$
$$+ k_{20} \cdot (1 - S_3 - S_4 - S_5 - S_6 - S_7)$$

$$\frac{dS_5}{dt} = k_{15} \cdot \frac{kn_6 \cdot S_1 \cdot S_4}{kn_6 + S_1} + k_{16} \cdot \frac{S_1 \cdot S_4}{kn_6 + S_1} - k_{18} \cdot \frac{kn_4 \cdot S_5}{kn_4 + S_1} - k_{19} \cdot \frac{kn_4 \cdot S_5}{kn_4 + S_1}$$
$$- k_{21} \cdot \frac{kn_4 \cdot S_1 \cdot S_5}{kn_4 + S_1} - k_{22} \cdot \frac{S_1 \cdot S_5}{kn_4 + S_1} + k_{23} \cdot S_7 + k_{24} \cdot S_7$$

$$\frac{dS_6}{dt} = k_{13} \cdot \frac{kn_4 \cdot kn_5 \cdot S_1 \cdot S_3}{kn_4 \cdot kn_5 + S_1 \cdot (kn_4 + kn_5)} + k_{14} \cdot \frac{kn_5 \cdot S_1 \cdot S_3}{kn_4 \cdot kn_5 + S_1 \cdot (kn_4 + kn_5)}$$
$$- k_{11} \cdot S_6 - k_{12} \cdot S_6$$

$$\frac{dS_7}{dt} = k_{21} \cdot \frac{kn_4 \cdot S_1 \cdot S_5}{kn_4 + S_1} + k_{22} \cdot \frac{S_1 \cdot S_5}{kn_4 + S_1} - k_{23} \cdot S_7 - k_{24} \cdot S_7$$

Notations and parameter values employed in Sneyd et al. [2004] are indicated in the following table.

S_1	C		k_8	l_{-4}	$2.5\ \mu M^{-1}s^{-1}$	k_{22}	l_2	$1.7\ s^{-1}$
S_2	C_t		k_9	k_2	$37.4\ \mu M^{-1}s^{-1}$	k_{23}	k_{-1}	$0.04\ s^{-1}$
S_3	R		k_{10}	l_4	$1.7\ \mu M^{-1}s^{-1}$	k_{24}	l_{-2}	$0.8\ s^{-1}$

S_4	O		k_{11}	k_{-1}	$0.04\ s^{-1}$	kn_1	K_S	$0.18\ \mu M$
S_5	A		k_{12}	l_{-2}	$0.8\ s^{-1}$	kn_2	K_p	$0.42\ \mu M$
S_6	I_1		k_{13}	k_1	$0.64\ s^{-1}$	kn_3	p	10
S_7	I_2		k_{14}	l_2	$1.7\ s^{-1}$	kn_4	L_1	$0.12\ \mu M$
k_1	k_f	$0.96\ s^{-1}$	k_{15}	k_4	$4\ \mu M^{-1}s^{-1}$	kn_5	L_3	$0.025\ \mu M$
k_2	g_1	$0.002\ s^{-1}$	k_{16}	l_6	$4707\ s^{-1}$	kn_6	L_5	$54.7\ \mu M$
k_3	V_s	$120\ \mu M^2 s^{-1}$	k_{17}	k_3	$0.11\ \mu M s^{-1}$	kn_7	$1/\gamma$	$1/5.4$
k_4	$\delta \cdot \alpha_1$	$0.01\text{-}0.03\ \mu M s^{-1}$	k_{18}	k_{-4}	$0.54\ s^{-1}$	n_1	n	4
k_5	$\delta \cdot \alpha_2$	$0.01\text{-}0.2\ s^{-1},$	k_{19}	l_{-6}	$11.4\ s^{-1}$	n_2	m	2
k_6	$\delta \cdot V_p$	$0.01\cdot 28\ \mu M s^{-1}$	k_{20}	k_{-3}	$29.8\ s^{-1}$			
k_7	k_{-2}	$1.4\ s^{-1}$	k_{21}	k_1	$0.64\ s^{-1}$			

De Young and Keizer [1992]: Closed cell calcium oscillations

The models proposed in De Young and Keizer [1992] focuses on intracellular calcium oscillations relying on the change in loci of the calcium from the ER to the cytosol through IP$_3$-receptor/channels. Influx and efflux to or from the cell are neglected, i.e. the sum of cytosolic and endoplasmic calcium is constant. In this thesis, the full model is used where IP$_3$ enters as a constant and is represented as nl-parameter. The IP$_3$-receptor/channel is assumed to have three subunits with three binding sites each, one for IP$_3$ and two for calcium (one of which for activation, one for inhibition). Cytosolic calcium is modeled as variable. The probability to yield a certain state for each of the three subunits is considered to be the same, and the 8 possible states of the IP$_3$-receptor/channel are included as variables. Since the values of receptor state variables are to be interpreted as probabilities, they sum up to one ($\sum_{i=2}^{9} S_i = 1$). Only the state where IP$_3$ and a calcium molecule is bound at the activation site (S_6) for all three subunits leads to influx of calcium into the cytosol through the IP$_3$-receptor channel. The authors conclude that the open probability of the calcium channel is proportional to S_6^3 (i.e. $n_1 = 3$). There is one calcium flux from ER to cytosol depending only on the concentration difference between the two compartments ("leak flux"), and the other flux through the IP$_3$-receptor/channels depending additionally on S_6^3. Calcium flux from cytosol to ER is assumed to follow Hill-type kinetics and reflects the work of ATP-dependent pumps. The model consists of 9 variables, 27 rate coefficients, 4 nl-parameters and 1 cooperativity parameter.

$$\frac{\mathrm{d}S_1}{\mathrm{d}t} = k_1 \cdot S_6^{n_1} \cdot (kn_3 - S_1 - kn_4 \cdot S_1) + k_2 \cdot (kn_3 - S_1 - kn_4 \cdot S_1) - \frac{k_3 \cdot S_1^{n_2}}{S_1^{n_2} + kn_1^{n_2}}$$

$$\frac{\mathrm{d}S_2}{\mathrm{d}t} = k_5 \cdot S_3 + k_{21} \cdot S_4 + k_{17} \cdot S_5 - (k_4 \cdot kn_2 + k_{16} \cdot S_1 + k_{20} \cdot S_1) \cdot S_2$$

$$\frac{\mathrm{d}S_3}{\mathrm{d}t} = k_4 \cdot kn_2 \cdot S_2 + k_{23} \cdot S_6 + k_9 \cdot S_7 - (k_8 \cdot S_1 + k_5 + k_{22} \cdot S_1) \cdot S_3$$

$$\frac{\mathrm{d}S_4}{\mathrm{d}t} = k_{20} \cdot S_1 \cdot S_2 + k_7 \cdot S_6 + k_{19} \cdot S_8 - (k_{18} \cdot S_1 + k_6 \cdot kn_2 + k_{21}) \cdot S_4$$

$$\frac{\mathrm{d}S_5}{\mathrm{d}t} = k_{16} \cdot S_1 \cdot S_2 + k_{13} \cdot S_7 + k_{25} \cdot S_8 - (k_{12} \cdot kn_2 + k_{17} + k_{24} \cdot S_1) \cdot S_5$$

$$\frac{\mathrm{d}S_6}{\mathrm{d}t} = k_{22} \cdot S_1 \cdot S_3 + k_6 \cdot kn_2 \cdot S_4 + k_{11} \cdot S_9 - (k_{10} \cdot S_1 + k_7 + k_{23}) \cdot S_6$$

$$\frac{\mathrm{d}S_7}{\mathrm{d}t} = k_8 \cdot S_1 \cdot S_3 + k_{12} \cdot kn_2 \cdot S_5 + k_{27} \cdot S_9 - (k_9 + k_{13} + k_{26} \cdot S_1) \cdot S_7$$

$$\frac{\mathrm{d}S_8}{\mathrm{d}t} = k_{18} \cdot S_1 \cdot S_4 + k_{24} \cdot S_1 \cdot S_5 + k_{15} \cdot S_9 - (k_{14} \cdot kn_2 + k_{19} + k_{25}) \cdot S_8$$

$$\frac{\mathrm{d}S_9}{\mathrm{d}t} = k_{10} \cdot S_1 \cdot S_6 + k_{26} \cdot S_1 \cdot S_7 + k_{14} \cdot kn_2 \cdot S_8 - (k_{15} + k_{11} + k_{27}) \cdot S_9$$

The notation used in De Young and Keizer [1992] is given in the following table. For IP$_3$, a concentration of 0.5 μM is used for determining the sensitivity at the reference parameter set.

S_1	$[Ca_i^{2+}]$		k_6	a_1	$400\ \mu M^{-1}s^{-1}$	k_{20}	a_5	$20\ \mu M^{-1}s^{-1}$
S_2	S_{000}		k_7	b_1	$52\ s^{-1}$	k_{21}	b_5	$1.6468\ s^{-1}$
S_3	S_{100}		k_8	a_2	$0.2\ \mu M^{-1}s^{-1}$	k_{22}	a_5	$20\ \mu M^{-1}s^{-1}$
S_4	S_{010}		k_9	b_2	$0.2098\ s^{-1}$	k_{23}	b_5	$1.6468\ s^{-1}$
S_5	S_{001}		k_{10}	a_2	$0.2\ \mu M^{-1}s^{-1}$	k_{24}	a_5	$20\ \mu M^{-1}s^{-1}$
S_6	S_{110}		k_{11}	b_2	$0.2098\ s^{-1}$	k_{25}	b_5	$1.6468\ s^{-1}$
S_7	S_{101}		k_{12}	a_3	$400\ \mu M^{-1}s^{-1}$	k_{26}	a_5	$20\ \mu M^{-1}s^{-1}$
S_8	S_{011}		k_{13}	b_3	$377.36\ s^{-1}$	k_{27}	b_5	$1.6468\ s^{-1}$
S_9	S_{111}		k_{14}	a_3	$400\ \mu M^{-1}s^{-1}$	kn_1	k_3	$0.1\ \mu M$
k_1	ν_1	$6\ s^{-1}$	k_{15}	b_3	$377.36\ s^{-1}$	kn_2	$[IP3]$	$(0.37\mu M, 0.62\mu M)$
k_2	ν_2	$0.11\ s^{-1}$	k_{16}	a_4	$0.2\ \mu M^{-1}s^{-1}$	kn_3	c_0	$2.0\ \mu M$
k_3	ν_3	$0.9\ \mu M^{-1}s^{-1}$	k_{17}	b_4	$0.0289\ s^{-1}$	kn_4	c_1	0.185

k_4	a_1	$400\ \mu M^{-1}s^{-1}$	k_{18}	a_4	$0.2\ \mu M^{-1}s^{-1}$	n_1	3
k_5	b_1	$52\ s^{-1}$	k_{19}	b_4	$0.0289\ s^{-1}$	n_2	2

A.7.2. Chain models of length four with mass action kinetics

The chain models of length four have been introduced in Wolf et al. [2005]. They are prototypes of oscillator models yielding oscillations for both a single positive and a single negative feedback. 4 species are modeled $(S_1, \ldots, S_4)$ which are connected by 8 reactions $(\nu_1, \ldots, \nu_8)$ and hence 8 according rate coefficients $(k_1, \ldots, k_8)$, 1 nl-parameter (kn_1) and 1 cooperativity parameter (n_1) which both tune the feedback term fb.

$$\frac{\mathrm{d}S_1}{\mathrm{d}t} = k_1 - k_2 \cdot S_1 \cdot fb - k_3 \cdot S_1$$
$$\frac{\mathrm{d}S_2}{\mathrm{d}t} = k_2 \cdot S_1 \cdot fb - k_4 \cdot S_2 - k_5 \cdot S_2$$
$$\frac{\mathrm{d}S_3}{\mathrm{d}t} = k_4 \cdot S_2 - k_6 \cdot S_3 - k_7 \cdot S_3$$
$$\frac{\mathrm{d}S_4}{\mathrm{d}t} = k_6 \cdot S_3 - k_8 \cdot S_4$$

The feedback term fb enters only in reaction ν_2 and is given by

$$fb = \begin{cases} \frac{kn_1^{n_1}}{kn_1^{n_1}+S_4^{n_1}} & \text{for the negative feedback} \\ \frac{kn_1^{n_1}+S_4^{n_1}}{kn_1^{n_1}} & \text{for the positive feedback.} \end{cases}$$

Only in the Appendix, section A.2.2, the positive feedback term has the form

$$fb = \frac{S_4^{n_1}}{kn_1^{n_1} + S_4^{n_1}}.$$

The reaction rates are defined by $\nu_1 = k_1$, $\nu_2 = k_2 \cdot S_1 \cdot fb$, $\nu_3 = k_3 \cdot S_1$, $\nu_4 = k_4 \cdot S_2$, $\nu_5 = k_5 \cdot S_2$, $\nu_6 = k_6 \cdot S_3$, $\nu_7 = k_7 \cdot S_3$, $\nu_8 = k_8 \cdot S_4$.

A.7.3. Chain models with different matter flow properties

The reaction rates are similar to those given in the chain model with matter flow between species and mass action kinetics. The model including matter flow in ν_2, ν_4 and ν_6 is exactly the same as given above.

matter flow in ν_2, ν_4, ν_6 (model C246)

$$\frac{dS_1}{dt} = \nu_1 - \nu_2 - \nu_3$$
$$\frac{dS_2}{dt} = \nu_2 - \nu_4 - \nu_5$$
$$\frac{dS_3}{dt} = \nu_4 - \nu_6 - \nu_7$$
$$\frac{dS_4}{dt} = \nu_6 - \nu_8$$

matter flow in ν_4, ν_6 (model C-46)

$$\frac{dS_1}{dt} = \nu_1 - \nu_3$$
$$\frac{dS_2}{dt} = \nu_2 - \nu_4 - \nu_5$$
$$\frac{dS_3}{dt} = \nu_4 - \nu_6 - \nu_7$$
$$\frac{dS_4}{dt} = \nu_6 - \nu_8$$

matter flow in ν_2, ν_4 (model C24-)

$$\frac{dS_1}{dt} = \nu_1 - \nu_2 - \nu_3$$
$$\frac{dS_2}{dt} = \nu_2 - \nu_4 - \nu_5$$
$$\frac{dS_3}{dt} = \nu_4 - \nu_7$$
$$\frac{dS_4}{dt} = \nu_6 - \nu_8$$

matter flow in ν_4 (model C-4-)

$$\frac{dS_1}{dt} = \nu_1 - \nu_3$$
$$\frac{dS_2}{dt} = \nu_2 - \nu_4 - \nu_5$$
$$\frac{dS_3}{dt} = \nu_4 - \nu_7$$
$$\frac{dS_4}{dt} = \nu_6 - \nu_8$$

matter flow in ν_2, ν_6 (model C2-6)

$$\frac{dS_1}{dt} = \nu_1 - \nu_2 - \nu_3$$
$$\frac{dS_2}{dt} = \nu_2 - \nu_5$$
$$\frac{dS_3}{dt} = \nu_4 - \nu_6 - \nu_7$$
$$\frac{dS_4}{dt} = \nu_6 - \nu_8$$

matter flow in ν_6 (model C--6)

$$\frac{dS_1}{dt} = \nu_1 - \nu_3$$
$$\frac{dS_2}{dt} = \nu_2 - \nu_5$$
$$\frac{dS_3}{dt} = \nu_4 - \nu_6 - \nu_7$$
$$\frac{dS_4}{dt} = \nu_6 - \nu_8$$

matter flow in ν_2 (model C2--)

$$\frac{dS_1}{dt} = \nu_1 - \nu_2 - \nu_3$$
$$\frac{dS_2}{dt} = \nu_2 - \nu_5$$
$$\frac{dS_3}{dt} = \nu_4 - \nu_7$$
$$\frac{dS_4}{dt} = \nu_6 - \nu_8$$

no matter flow (model C---)

$$\frac{dS_1}{dt} = \nu_1 - \nu_3$$
$$\frac{dS_2}{dt} = \nu_2 - \nu_5$$
$$\frac{dS_3}{dt} = \nu_4 - \nu_7$$
$$\frac{dS_4}{dt} = \nu_6 - \nu_8$$

A.7.4. Chain models with saturating kinetics

All reactions except for the constant production ν_1 are modeled by Michaelis-Menten kinetics. There are 4 species $(S_1, \ldots, S_4)$, 8 rate coefficients (referred to as V_i,

$i = 2, \ldots, 8$), 8 nl-parameters $(kn_1, K_2, \ldots, K_8)$ and 1 cooperativity parameter (n_1).

$$\frac{\mathrm{d}S_1}{\mathrm{d}t} = k_1 - V_2 \cdot \frac{S_1}{S_1 + K_2} \cdot fb - V_3 \cdot \frac{S_1}{S_1 + K_3}$$

$$\frac{\mathrm{d}S_2}{\mathrm{d}t} = V_2 \cdot \frac{S_1}{S_1 + K_2} \cdot fb - V_4 \cdot \frac{S_2}{S_2 + K_4} - V_5 \cdot \frac{S_2}{S_2 + K_5}$$

$$\frac{\mathrm{d}S_3}{\mathrm{d}t} = V_4 \cdot \frac{S_2}{S_2 + K_4} - V_6 \cdot \frac{S_3}{S_3 + K_6} - V_7 \cdot \frac{S_3}{S_3 + K_7}$$

$$\frac{\mathrm{d}S_4}{\mathrm{d}t} = V_6 \cdot \frac{S_3}{S_3 + K_6} - V_8 \cdot \frac{S_4}{S_4 + K_8}$$

A.7.5. Chain models of length five

These models are similar to the chain models of length four including one more species S_5. fb denotes the feedback term from the chain models of length four. 5 species, 10 rate coefficients, 1 nl-parameter and 1 cooperativity parameter are modeled.

$$\frac{\mathrm{d}S_1}{\mathrm{d}t} = k_1 - k_2 \cdot S_1 \cdot fb - k_3 \cdot S_1$$

$$\frac{\mathrm{d}S_2}{\mathrm{d}t} = k_2 \cdot S_1 \cdot fb - k_4 \cdot S_2 - k_5 \cdot S_2$$

$$\frac{\mathrm{d}S_3}{\mathrm{d}t} = k_4 \cdot S_2 - k_6 \cdot S_3 - k_7 \cdot S_3$$

$$\frac{\mathrm{d}S_4}{\mathrm{d}t} = k_6 \cdot S_3 - k_8 \cdot S_4 - k_9 \cdot S_4$$

$$\frac{\mathrm{d}S_5}{\mathrm{d}t} = k_8 \cdot S_4 - k_{10} \cdot S_5$$

A.7.6. Models according to Tsai et al. [2008] and alterations

The models proposed in Tsai et al. [2008] are based on three species (or protein) conversion cycles where the sums of the two species in a conversion cycle are constant $(S_i + S_i^* = 1, \; i = 1, 2, 3)$. The conversion cycles are circularly connected by one activating regulation from S_i to another cycle. Thereby, S_i positively regulates a conversion S_j to S_j^*, i.e. the conversion from the regulatory species to the non-regulatory species. S_1 regulates S_2, S_2 regulates S_3 and S_3 regulates S_1. In the model the ODE systems are simplified by using the conditions $S_i^* = 1 - S_i$ and thus replacing the ODEs for S_i^*, $i = 1, 2, 3$. The model consists of 3 species, 6 rate coefficients, 3 nl-parameters and 3 cooperativity parameters.

Model with negative feedback only

$$\frac{\mathrm{d}S_1}{\mathrm{d}t} = k_1 \cdot (1 - S_1) - k_2 \cdot S_1 \cdot \frac{S_3^{n_1}}{S_3^{n_1} + kn_1^{n_1}}$$

$$\frac{\mathrm{d}S_2}{\mathrm{d}t} = k_3 \cdot (1 - S_2) - k_4 \cdot S_2 \cdot \frac{S_1^{n_2}}{S_1^{n_2} + kn_2^{n_2}}$$

$$\frac{\mathrm{d}S_3}{\mathrm{d}t} = k_5 \cdot (1 - S_3) - k_6 \cdot S_3 \cdot \frac{S_2^{n_3}}{S_2^{n_3} + kn_3^{n_3}}$$

Model with positive feedback

In this model also proposed in Tsai et al. [2008], an additional conversion from S_1^* to S_1 is included which is positively regulated by S_1 (denoted here by reaction ν_7). The model consists of 3 species, 7 rate coefficients, 4 nl-parameters and 4 cooperativity parameters.

$$\frac{\mathrm{d}S_1}{\mathrm{d}t} = k_1 \cdot (1 - S_1) - k_2 \cdot S_1 \cdot \frac{S_3^{n_1}}{S_3^{n_1} + kn_1^{n_1}} + k_7 \cdot (1 - S_1) \cdot \frac{S_1^{n_4}}{S_1^{n_4} + kn_4^{n_4}}$$

$$\frac{\mathrm{d}S_2}{\mathrm{d}t} = k_3 \cdot (1 - S_2) - k_4 \cdot S_2 \cdot \frac{S_1^{n_2}}{S_1^{n_2} + kn_2^{n_2}}$$

$$\frac{\mathrm{d}S_3}{\mathrm{d}t} = k_5 \cdot (1 - S_3) - k_6 \cdot S_3 \cdot \frac{S_2^{n_3}}{S_2^{n_3} + kn_3^{n_3}}$$

In the publication (Tsai et al. [2008]), the following notations and ranges or values for the parameters are used. Note that the cooperativity parameters are varied in Tsai et al. [2008] but are are fixed to 3 in the examination in this thesis since the negative feedback only model displays no oscillations for smaller Hill coefficients (not shown). The different intervals for k_7 refer to the model with negative feedback only, the model with a weak positive feedback and the model with a stronger positive feedback, respectively.

S_1	A		k_4	k_4	[0,1000]	kn_3	K_3	[0,4]
S_2	B		k_5	k_5	1	kn_4	K_5	[0,4]
S_3	C		k_6	k_6	[0,1000]	n_1	n_1	[1,4]
k_1	k_1	[0,10]	k_7	k_7	0,[0,100],[500,600]	n_2	n_2	[1,4]
k_2	k_2	[0,1000]	kn_1	K_1	[0,4]	n_3	n_3	[1,4]
k_3	k_3	[0,10]	kn_2	K_2	[0,4]	n_4	n_4	[1,4]

Model with negative feedback and Michaelis-Menten kinetics for degradations

This model is an alteration of the model proposed in Tsai et al. [2008] and is used in this thesis to examine the effects of saturating kinetics on period and amplitude sensitivity. Thereby, the negative feedback-only model proposed in Tsai et al. [2008] is used as basis and saturating kinetics are employed for the conversions from S_i to S_i^*, $i = 1, 2, 3$ (reactions ν_2, ν_4, ν_6). The model consists of 3 species, 6 rate coefficients, 7 nl-parameters and 4 cooperativity parameters.

$$\frac{\mathrm{d}S_1}{\mathrm{d}t} = k_1 \cdot (1 - S_1) - k_2 \cdot \frac{S_1}{S_1 + kn_4} \cdot \frac{S_3^{n_1}}{S_3^{n_1} + kn_1^{n_1}}$$
$$\frac{\mathrm{d}S_2}{\mathrm{d}t} = k_3 \cdot (1 - S_2) - k_4 \cdot \frac{S_2}{S_2 + kn_5} \cdot \frac{S_1^{n_2}}{S_1^{n_2} + kn_2^{n_2}}$$
$$\frac{\mathrm{d}S_3}{\mathrm{d}t} = k_5 \cdot (1 - S_3) - k_6 \cdot \frac{S_3}{S_3 + kn_6} \cdot \frac{S_2^{n_3}}{S_2^{n_3} + kn_3^{n_3}}$$

A.7.7. Synthetic oscillators

Repressilator - original model

The repressilator proposed in Elowitz and Leibler [2000] is constructed by three gene-protein modules using the genes *lacI* of *E.coli*, *tetR* (tetracyclin resistance) and *cI* from λ phage. In each module, the mRNA is transcribed and degraded and the protein is translated from the mRNA and also degraded. The proteins repress the transcription of one other mRNA such that a circular regulation results. 6 variables, 15 rate coefficients, 3 nl-parameters and 3 cooperativity parameters are employed in the model.

$$\frac{dS_1}{dt} = -k_1 \cdot S_1 + k_2 \cdot S_4$$

$$\frac{dS_2}{dt} = -k_3 \cdot S_2 + k_4 \cdot S_5$$

$$\frac{dS_3}{dt} = -k_5 \cdot S_3 + k_6 \cdot S_6$$

$$\frac{dS_4}{dt} = k_7 + k_8 \cdot \frac{kn_1^{n_1}}{kn_1^{n_1} + S_3^{n_1}} - k_9 \cdot S_4$$

$$\frac{dS_5}{dt} = k_{10} + k_{11} \cdot \frac{kn_2^{n_2}}{kn_2^{n_2} + S_1^{n_2}} - k_{12} \cdot S_5$$

$$\frac{dS_6}{dt} = k_{13} + k_{14} \cdot \frac{kn_3^{n_3}}{kn_3^{n_3} + S_2^{n_3}} - k_{15} \cdot S_6$$

In the original repressilator model as proposed in Elowitz and Leibler [2000], the following notations and values for the parameters are used.

S_1	o_1		k_4	β	0.2	k_{13}	α	$5 \cdot 10^{-4}$
S_2	o_2		k_5	β/K_M	8	k_{14}	k_{-1}	$0.5 - 5 \cdot 10^{-4}$
S_3	o_3		k_6	β	0.2	k_{15}		0.05
S_4	m_1		k_7	α_0	$5 \cdot 10^{-4}$	kn_1	K_M	40
S_5	m_2		k_8	α	$0.5 - 5 \cdot 10^{-4}$	kn_2	K_M	40
S_6	m_3		k_9		0.05	kn_3	K_M	40
k_1	β/K_M	8	k_{10}	α_0	$5 \cdot 10^{-4}$	n_1	n	2
k_2	β	0.2	k_{11}	α	$0.5 - 5 \cdot 10^{-4}$	n_2	n	2
k_3	β/K_M	8	k_{12}		0.05	n_3	n	2

Repressilator with saturating kinetics

To examine the effect of saturating kinetics, the linear protein productions and the linear protein and mRNA degradations are replaced by Michaelis-Menten terms. The model yields still 6 variables, 15 rate coefficients, and 3 cooperativity parameters, but now 12 nl-parameters.

$$\frac{dS_1}{dt} = -k_1 \cdot \frac{S_1}{S_1 + kn_1} + k_2 \cdot \frac{S_4}{S_4 + kn_2}$$

$$\frac{dS_2}{dt} = -k_3 \cdot \frac{S_2}{S_2 + kn_3} + k_4 \cdot \frac{S_5}{S_5 + kn_4}$$

$$\frac{dS_3}{dt} = -k_5 \cdot \frac{S_3}{S_3 + kn_5} + k_6 \cdot \frac{S_6}{S_6 + kn_6}$$

$$\frac{dS_4}{dt} = k_7 + k_8 \cdot \frac{kn_7^{n_1}}{kn_7^{n_1} + S_3^{m_1}} - k_9 \cdot \frac{S_4}{S_4 + kn_8}$$

$$\frac{dS_5}{dt} = k_{10} + k_{11} \cdot \frac{kn_9^{n_2}}{kn_9^{n_2} + S_1^{n_2}} - k_{12} \cdot \frac{S_5}{S_5 + kn_{10}}$$

$$\frac{dS_6}{dt} = k_{13} + k_{14} \cdot \frac{kn_{11}^{n_3}}{kn_{11}^{n_3} + S_2^{n_3}} - k_{15} \cdot \frac{S_6}{S_6 + kn_{12}}$$

Metabolator - original model

The metabolator has been proposed by Fung et al. [2005] who have implemented it in *E. coli*. It combines transcriptional regulation and metabolism. In particular, the acetate pathway is used in which the acetyl coenzyme A is converted forth and back to the pool of acetyl phosphate, acetate and protonated acetate. This pool induces production of the proteins LacI and Acs. Acs promotes conversion of the pool to acetyl coenzyme A, LacI represses the production of the protein Pta which induces the conversion from acetyl coenzyme A to the other pool. It incorporates 7 variables, 18 rate coefficients, 5 nl-parameters and 3 cooperativity parameters.

$$\frac{dS_1}{dt} = k_1 \cdot S_6 \cdot \frac{S_3}{kn_1 + S_3} - k_2 \cdot S_7 \cdot \frac{S_1}{kn_2 + S_1} + k_3 - k_4 \cdot S_1$$

$$\frac{dS_2}{dt} = k_2 \cdot S_7 \cdot \frac{S_1}{kn_2 + S_1} - k_5 \cdot S_2 + k_6 \cdot S_3$$

$$\frac{dS_3}{dt} = k_5 \cdot S_2 - k_6 \cdot S_3 - k_7 \cdot S_3 + k_8 \cdot S_4 - k_1 \cdot S_6 \cdot \frac{S_3}{kn_1 + S_3}$$

$$\frac{dS_4}{dt} = k_7 \cdot S_3 - k_8 \cdot S_4 - k_9 \cdot S_4$$

$$\frac{dS_5}{dt} = k_{10} \cdot \frac{S_2^{n_1}}{kn_3^{n_1} + S_2^{n_1}} + k_{11} - k_{12} \cdot S_5$$

$$\frac{dS_6}{dt} = k_{13} \cdot \frac{S_2^{n_2}}{kn_4^{n_2} + S_2^{n_2}} + k_{14} - k_{15} \cdot S_6$$

$$\frac{dS_7}{dt} = k_{16} \cdot \frac{kn_5^{n_3}}{kn_5^{n_3} + S_5^{n_3}} + k_{17} - k_{18} \cdot S_7$$

In the original model, the following notation and parameter values are employed.

S_1	AcCoA		k_5	$k_{Ack,f}$	1	k_{16}	α_3	2
S_2	AcP		k_6	$k_{Ack,r}$	1	k_{17}	α_0	0
S_3	OAc$^-$		k_7	$C \cdot H^+$	$100 \cdot 10^{-7}$	k_{18}	k_d	0.06
S_4	HOAc		k_8	$C \cdot K_{eq}$	$100 \cdot 10^{-4.5}$	kn_1	$K_{m,2}$	0.1
S_5	LacI		k_9	k_3	0.01	kn_2	$K_{m,1}$	0.06
S_6	Acs		k_{10}	α_1	0.1	kn_3	$K_{g,1}$	10
S_7	Pta		k_{11}	α_0	0	kn_4	$K_{g,2}$	10
k_1	k_2	0.8	k_{12}	k_d	0.06	kn_5	$K_{g,3}$	0.001
k_2	k_1	80	k_{13}	α_2	2	n_1	n	2
k_3	S_0	0.5	k_{14}	α_0	0	n_2	n	2
k_4	k_{TCA}	10	k_{15}	k_d	0.06	n_3	n	2

Metabolator with saturating kinetics

In order to examine the effects of saturating kinetics, all linear reactions are replaced by Michaelis-Menten kinetics reactions. The model incorporates still 7 variables, 18 rate coefficients, and 3 cooperativity parameters, but 16 nl-parameters.

$$\frac{dS_1}{dt} = k_1 \cdot \frac{S_6}{kn_2 + S_6} \cdot \frac{S_3}{kn_1 + S_3} - k_2 \cdot \frac{S_7}{kn_4 + S_7} \cdot \frac{S_1}{kn_3 + S_1} + k_3 - k_4 \cdot \frac{S_1}{kn_5 + S_1}$$

$$\frac{dS_2}{dt} = k_2 \cdot \frac{S_7}{kn_4 + S_7} \cdot \frac{S_1}{kn_3 + S_1} - k_5 \cdot \frac{S_2}{kn_6 + S_2} + k_6 \cdot \frac{S_3}{kn_7 + S_3}$$

$$\frac{dS_3}{dt} = k_5 \cdot \frac{S_2}{kn_6 + S_2} - k_6 \cdot \frac{S_3}{kn_7 + S_3} - k_7 \cdot \frac{S_3}{kn_8 + S_3} + k_8 \cdot \frac{S_4}{kn_9 + S_4}$$
$$\quad - k_1 \cdot \frac{S_6}{kn_2 + S_6} \cdot \frac{S_3}{kn_1 + S_3}$$

$$\frac{dS_4}{dt} = k_7 \cdot \frac{S_3}{kn_8 + S_3} - k_8 \cdot \frac{S_4}{kn_9 + S_4} - k_9 \cdot \frac{S_4}{kn_{10} + S_4}$$

$$\frac{dS_5}{dt} = k_{10} \cdot \frac{S_2^{m_1}}{kn_{11}^{n_1} + S_2^{n_1}} + k_{11} - k_{12} \cdot \frac{S_5}{kn_{12} + S_5}$$

$$\frac{dS_6}{dt} = k_{13} \cdot \frac{S_2^{m_2}}{kn_{13}^{n_2} + S_2^{n_2}} + k_{14} - k_{15} \cdot \frac{S_6}{kn_{14} + S_6}$$

$$\frac{dS_7}{dt} = k_{16} \cdot \frac{kn_{15}^{n_3}}{kn_{15}^{n_3} + S_5^{n_3}} + k_{17} - k_{18} \cdot \frac{S_7}{kn_{16} + S_7}$$

A.7.8. NF-κB oscillations model

The model proposed by Kearns et al. [2006] is used as basis and it is reduced thereby keeping the principal dynamics according to a work by J. Mothes (to be published, personal communication). Modeled are nuclear and cytosolic NF-κB, IκBα mRNA, its cytosolic and nuclear IκBα concentrations, IKK and the concentrations of some complexes (see notation in the original model in the table below). IKK and NF-κB underly conservation relations. Nuclear NF-κB activates the production of IκBα mRNA, cytosolic IκBα sequesters NF-κB in the cytosol. Total IKK and total NF-κB concentrations are encoded as nl-parameters kn_1 and kn_2, respectively. The reduced model consists of 8 variables, 16 rate coefficients, 2 nl-parameters and 1 cooperativity parameter.

$$\frac{dS_1}{dt} = -k_1 \cdot S_1 \cdot S_4 - k_2 \cdot S_1 \cdot (kn_1 - S_6 - S_7) + k_3 \cdot S_7 - k_4 \cdot S_1$$

$$\frac{dS_2}{dt} = k_4 \cdot S_1 - k_5 \cdot S_2 \cdot S_5$$

$$\frac{dS_3}{dt} = k_6 + k_7 \cdot S_2^{n_1} - k_8 \cdot S_3$$

$$\frac{dS_4}{dt} = -k_1 \cdot S_1 \cdot S_4 - k_9 \cdot S_4 \cdot S_6 + k_{10} \cdot (kn_1 - S_6 - S_7) + k_{11} \cdot S_3 - k_{12} \cdot S_4 - k_{13} \cdot S_4$$

$$\frac{dS_5}{dt} = -k_5 \cdot S_2 \cdot S_5 + k_{13} \cdot S_4 - k_{15} \cdot S_5$$

$$\frac{dS_6}{dt} = k_3 \cdot S_7 - k_9 \cdot S_4 \cdot S_6 + k_{10} \cdot (kn_1 - S_6 - S_7) - k_{14} \cdot S_6 \cdot S_8$$

$$\frac{dS_7}{dt} = k_2 \cdot S_1 \cdot (kn_1 - S_6 - S_7) - k_3 \cdot S_7 + k_{14} \cdot S_6 \cdot S_8$$

$$\frac{dS_8}{dt} = k_1 \cdot S_1 \cdot S_4 + k_{16} \cdot (kn_2 - S_8 - S_7 - S_1 - S_2) - k_{14} \cdot S_6 \cdot S_8$$

The system is modeled under stimulated conditions (increased IKK_{tot} value). Notation which will be used by J. Mothes in the publication, the parameter values and calculated stimulated steady state values (which can used slightly perturbed as initial conditions, in μM) are given in the following. Units are min^{-1} or $min^{-1}\mu M^{-1}$ for the rate coefficients, and μM for the nl-parameters.

| S_1 | $NF\kappa B$ | 0.0032 | k_2 | $a4i$ | 30 | k_{11} | $tr1a$ | 0.2448 |
| S_2 | $NF\kappa B_n$ | 0.1605 | k_3 | $r4$ | 0.36 | k_{12} | $deg1$ | 0.12 |

S_3	$I\kappa B\alpha_t$	1.0677	k_4	$k1$	5.4	k_{13}	$tp1a$	0.018
S_4	$I\kappa B\alpha$	0.9758	k_5	$a4n$	30	kn_{14}	$a7$	11.1
S_5	$I\kappa B\alpha_n$	0.0036	k_6	$tr2a$	$1.848 \cdot 10^{-4}$	kn_{15}	$deg1n$	0.12
S_6	IKK	0.0459	k_7	$tr4a$	1.386	kn_{16}	$k2a$	0.828
S_7	$I\kappa B\alpha NF\kappa BIKK$	0.3996	k_8	$tr3a$	0.0336	kn_1	IKK_{tot}	0.80056
S_8	$I\kappa B\alpha NF\kappa B$	0.2161	k_9	$a1$	1.35	kn_2	$NF\kappa B_{tot}$	0.8
k_1	$a4$	30	k_{10}	$d1$	0.075	n_1		2

Bibliography

K. R. Albe, M. H. Butler, and B. E. Wright. Cellular concentrations of enzymes and their substrates. *Journal of Theoretical Biology*, 143(2):163–195, 1990.

M. Apri, J. Molenaar, M. de Gee, and G. van Voorn. Efficient estimation of the robustness region of biological models with oscillatory behavior. *PLoS One*, 5(4):e9865, 2010.

B. D. Aronson, K. A. Johnson, J. J. Loros, and J. C. Dunlap. Negative feedback defining a circadian clock - autoregulation of the clock gene-frequency. *Science*, 263(5153):1578–1584, 1994.

L. Ashall, C. A. Horton, D. E. Nelson, P. Paszek, C. V. Harper, K. Sillitoe, S. Ryan, D. G. Spiller, J. F. Unitt, D. S. Broomhead, D. B. Kell, D. A. Rand, V. See, and M. R. White. Pulsatile stimulation determines timing and specificity of NF-kappaB-dependent transcription. *Science*, 324(5924):242–6, 2009.

J. E. Baggs, T. S. Price, L. DiTacchio, S. Panda, G. A. Fitzgerald, and J. B. Hogenesch. Network features of the mammalian circadian clock. *PLoS Biol*, 7(3):e52, 2009.

N. Bagheri, J. Stelling, and F. J. Doyle. Quantitative performance metrics for robustness in circadian rhythms. *Bioinformatics*, 23(3):358–364, 2007.

N. Bagheri, M. J. Lawson, J. Stelling, and 3rd Doyle, F. J. Modeling the Drosophila melanogaster circadian oscillator via phase optimization. *J Biol Rhythms*, 23(6):525–37, 2008.

N. Barkai and S. Leibler. Robustness in simple biochemical networks. *Nature*, 387(6636):913–7, 1997.

S. Becker-Weimann, J. Wolf, H. Herzel, and A. Kramer. Modeling feedback loops of the mammalian circadian oscillator. *Biophys J*, 87(5):3023–34, 2004.

M. J. Berridge and A. Galione. Cytosolic calcium oscillators. *FASEB journal : official publication of the Federation of American Societies for Experimental Biology*, 2(15):3074–82, 1988.

M. J. Berridge, P. Lipp, and M. D. Bootman. The versatility and universality of calcium signalling. *Nature Reviews Molecular Cell Biology*, 1(1):11–21, 2000.

N. Blüthgen. Sequestration shapes the response of signal transduction cascades. *IUBMB Life*, 58(11):659–63, 2006.

N. Blüthgen and S. Legewie. Systems analysis of MAPK signal transduction. *Essays in biochemistry*, 45:95–107, 2008.

N. Blüthgen, F. J. Bruggeman, S. Legewie, H. Herzel, H. V. Westerhoff, and B. N. Kholodenko. Effects of sequestration on signal transduction cascades. *FEBS J*, 273(5):895–906, 2006.

M. D. Bootman. Calcium signaling. *Cold Spring Harb Perspect Biol*, 4(7):a011171, 2012.

G. Bordyugov, P. O. Westermark, A. Korencic, S. Bernard, and H. Herzel. Mathematical modeling in chronobiology. *Handbook of experimental pharmacology*, 217:335–57, 2013.

J. A. M. Borghans, R. J. DeBoer, and L. A. Segel. Extending the quasi-steady state approximation by changing variables. *Bulletin of Mathematical Biology*, 58(1):43–63, 1996.

S. Boslaugh and P. A. Watters. *Statistics in a Nutshell*. O'Reilly Media, Inc., Sebastopol, 1 edition, 2008.

O. Brandman and T. Meyer. Feedback loops shape cellular signals in space and time. *Science*, 322(5900):390–5, 2008.

G. E. Briggs and J. B. Haldane. A note on the kinetics of enzyme action. *The Biochemical journal*, 19(2):338–9, 1925.

N. E. Buchler, U. Gerland, and T. Hwa. Nonlinear protein degradation and the function of genetic circuits. *Proc Natl Acad Sci U S A*, 102(27):9559–64, 2005.

L. Cai, C. K. Dalal, and M. B. Elowitz. Frequency-modulated nuclear localization bursts coordinate gene regulation. *Nature*, 455(7212):485–90, 2008.

P. Camacho and J. D. Lechleiter. Increased frequency of calcium waves in Xenopus laevis oocytes that express a calcium-ATPase. *Science*, 260(5105):226–9, 1993.

S. S. Chen, P. Harrigan, B. Heineike, J. Stewart-Ornstein, and H. El-Samad. Building robust functionality in synthetic circuits using engineered feedback regulation. *Current Opinion in Biotechnology*, 24(4):790–796, 2013.

P. Cheng, Y. Yang, and Y. Liu. Interlocked feedback loops contribute to the robustness of the Neurospora circadian clock. *Proc Natl Acad Sci U S A*, 98(13):7408–13, 2001.

A. Ciliberto, F. Capuani, and J. J. Tyson. Modeling networks of coupled enzymatic reactions using the total quasi-steady state approximation. *PLoS Comput Biol*, 3(3):e45, 2007.

N. A. Cookson, W. H. Mather, T. Danino, O. Mondragon-Palomino, R. J. Williams, L. S. Tsimring, and J. Hasty. Queueing up for enzymatic processing: correlated signaling through coupled degradation. *Molecular Systems Biology*, 7, 2011.

A. Cornish-Bowden. *Fundamentals of enzyme kinetics*. Portland Press Ltd, London, 3rd edition, 2004.

M. E. Csete and J. C. Doyle. Reverse engineering of biological complexity. *Science*, 295 (5560):1664–9, 2002.

A. Dayarian, M. Chaves, E. D. Sontag, and A. M. Sengupta. Shape, size, and robustness: feasible regions in the parameter space of biochemical networks. *PLoS Comput Biol*, 5 (1):e1000256, 2009.

P. De Koninck and H. Schulman. Sensitivity of CaM kinase II to the frequency of Ca2+ oscillations. *Science*, 279(5348):227–30, 1998.

G. W. De Young and J. Keizer. A single-pool inositol 1,4,5-trisphosphate-receptor-based model for agonist-stimulated oscillations in Ca2+ concentration. *Proc Natl Acad Sci U S A*, 89(20):9895–9, 1992.

O. Decroly and A. Goldbeter. Birhythmicity, chaos, and other patterns of temporal self-organization in a multiply regulated biochemical system. *Proc Natl Acad Sci U S A*, 79 (22):6917–21, 1982.

R. E. Dolmetsch, K. Xu, and R. S. Lewis. Calcium oscillations increase the efficiency and specificity of gene expression. *Nature*, 392(6679):933–6, 1998.

J. Dunlap. Circadian rhythms. An end in the beginning. *Science*, 280(5369):1548–9, 1998.

G. Dupont, L. Combettes, G. S. Bird, and J. W. Putney. Calcium oscillations. *Cold Spring Harb Perspect Biol*, 3(3), 2011.

M. B. Elowitz and S. Leibler. A synthetic oscillatory network of transcriptional regulators. *Nature*, 403(6767):335–8, 2000.

B. Ermentrout. *Simulating, analyzing, and animating dynamical systems: a guide to XP-PAUT for researchers and students*. Software, Environment, and Tools. siam, Philadelphia, 2002.

T. Evans, E. T. Rosenthal, J. Youngblom, D. Distel, and T. Hunt. Cyclin: a protein specified by maternal mRNA in sea urchin eggs that is destroyed at each cleavage division. *Cell*, 33(2):389–96, 1983.

D. Fell. *Understanding the control of metabolism*. Frontiers in metabolism 2. Portland Press Ltd, London, 1997.

Jr. Ferrell, J. E. Self-perpetuating states in signal transduction: positive feedback, double-negative feedback and bistability. *Curr Opin Cell Biol*, 14(2):140–8, 2002.

E. H. Flach and S. Schnell. Use and abuse of the quasi-steady-state approximation. *Syst Biol (Stevenage)*, 153(4):187–91, 2006.

R. Fritsche-Guenther, F. Witzel, A. Sieber, R. Herr, N. Schmidt, S. Braun, T. Brummer, C. Sers, and N. Blüthgen. Strong negative feedback from Erk to Raf confers robustness to MAPK signalling. *Molecular systems biology*, 7:489, 2011.

E. Fung, W. W. Wong, J. K. Suen, T. Bulter, S. G. Lee, and J. C. Liao. A synthetic gene-metabolic oscillator. *Nature*, 435(7038):118–22, 2005.

J. S. Geller and M. Brenner. Measurements of metabolites during cAMP oscillations of Dictyostelium discoideum. *Journal of cellular physiology*, 97(3 Pt 2 Suppl 1):413–20, 1978.

C. Gerard, D. Gonze, and A. Goldbeter. Dependence of the period on the rate of protein degradation in minimal models for circadian oscillations. *Philos Transact A Math Phys Eng Sci*, 367(1908):4665–83, 2009.

C. Gerard, D. Gonze, and A. Goldbeter. Effect of positive feedback loops on the robustness of oscillations in the network of cyclin-dependent kinases driving the mammalian cell cycle. *FEBS J*, 2012.

A. Ghosh and B. Chance. Oscillations of glycolytic intermediates in yeast cells. *Biochemical and Biophysical Research Communications*, 16(2):174–81, 1964.

J. D. Gibbons and S. Chakraborti. *Nonparametric statistical inference*. Statistics: Textbooks and Monographs. Chapman & Hall/CRC, Boca Raton, 5 edition, 2011.

A. Goldbeter. A minimal cascade model for the mitotic oscillator involving cyclin and cdc2 kinase. *Proceedings of the National Academy of Sciences of the United States of America*, 88(20):9107–11, 1991.

A. Goldbeter. A model for circadian oscillations in the Drosophila period protein (PER). *Proc Biol Sci*, 261(1362):319–24, 1995.

A. Goldbeter. Computational approaches to cellular rhythms. *Nature*, 420(6912):238–45, 2002.

A. Goldbeter. Oscillatory enzyme reactions and Michaelis-Menten kinetics. *FEBS letters*, 587(17):2778–84, 2013.

A. Goldbeter and Jr. Koshland, D. E. An amplified sensitivity arising from covalent modification in biological systems. *Proc Natl Acad Sci U S A*, 78(11):6840–4, 1981.

A. Goldbeter and Jr. Koshland, D. E. Ultrasensitivity in biochemical systems controlled by covalent modification. interplay between zero-order and multistep effects. *J Biol Chem*, 259(23):14441–7, 1984.

A. Goldbeter, G. Dupont, and M. J. Berridge. Minimal model for signal-induced Ca2+ oscillations and for their frequency encoding through protein phosphorylation. *Proc Natl Acad Sci U S A*, 87(4):1461–5, 1990.

A. Goldbeter, C. Gerard, D. Gonze, J. C. Leloup, and G. Dupont. Systems biology of cellular rhythms. *FEBS Lett*, 586(18):2955–65, 2012.

D. Gonze and M. Hafner. *Positive Feedbacks Contribute to the Robustness of the Cell Cycle with Respect to Molecular Noise*, volume 407 of *Lecture Notes in Control and Information Sciences*, chapter 23, pages 283–295. Springer Berlin Heidelberg, 2011.

D. Gonze, J. Halloy, and A. Goldbeter. Robustness of circadian rhythms with respect to molecular noise. *Proc Natl Acad Sci U S A*, 99(2):673–8, 2002.

D. Gonze, S. Bernard, C. Waltermann, A. Kramer, and H. Herzel. Spontaneous synchronization of coupled circadian oscillators. *Biophysical Journal*, 89(1):120–129, 2005.

D. Gonze, M. Jacquet, and A. Goldbeter. Stochastic modelling of nucleocytoplasmic oscillations of the transcription factor Msn2 in yeast. *J R Soc Interface*, 5 Suppl 1: S95–109, 2008.

B. C. Goodwin. Oscillatory behavior in enzymatic control processes. *Adv Enzyme Regul*, 3:425–38, 1965.

J. S. Griffith. Mathematics of cellular control processes .i. negative feedback to 1 gene. *Journal of Theoretical Biology*, 20(2):202–&, 1968a.

J. S. Griffith. Mathematics of cellular control processes. ii. positive feedback to one gene. *J Theor Biol*, 20(2):209–16, 1968b.

V. Grubelnik, A. Z. Larsen, U. Kummer, L. F. Olsen, and M. Marhl. Mitochondria regulate the amplitude of simple and complex calcium oscillations. *Biophysical Chemistry*, 94 (1-2):59–74, 2001.

T. Haberichter, M. Marhl, and R. Heinrich. Birhythmicity, trirhythmicity and chaos in bursting calcium oscillations. *Biophys Chem*, 90(1):17–30, 2001.

M. Hafner, H. Koeppl, M. Hasler, and A. Wagner. 'Glocal' robustness analysis and model discrimination for circadian oscillators. *PLoS Comput Biol*, 5(10):e1000534, 2009.

R. Heinrich and T. A. Rapoport. A linear steady-state treatment of enzymatic chains. general properties, control and effector strength. *Eur J Biochem*, 42(1):89–95, 1974.

J. Higgins. A chemical mechanism for oscillation of glycolytic intermediates in yeast cells. *Proceedings of the National Academy of Sciences of the United States of America*, 51: 989–94, 1964.

A. L. Hodgkin, A. F. Huxley, and B. Katz. Measurement of current-voltage relations in the membrane of the giant axon of Loligo. *The Journal of physiology*, 116(4):424–48, 1952.

J. B. Hogenesch and E. D. Herzog. Intracellular and intercellular processes determine robustness of the circadian clock. *FEBS Lett*, 585(10):1427–34, 2011.

J. B. Hogenesch and H. R. Ueda. Understanding systems-level properties: timely stories from the study of clocks. *Nat Rev Genet*, 12(6):407–16, 2011.

J.H. Hubbard and B.H. West. *Differential Equations, A Dynamical Systems Approach: Higher Dimensional Systems*, volume No. 18 of *Texts in Applied Mathematics*. Springer-Verlag, New York, 1995.

F. Hussain, C. Gupta, A. J. Hirning, W. Ott, K. S. Matthews, K. Josic, and M. R. Bennett. Engineered temperature compensation in a synthetic genetic clock. *Proceedings of the National Academy of Sciences of the United States of America*, 111(3):972–7, 2014.

A. E. Ihekwaba, D. S. Broomhead, R. L. Grimley, N. Benson, and D. B. Kell. Sensitivity analysis of parameters controlling oscillatory signalling in the NF-kappaB pathway: the roles of IKK and IkappaBalpha. *Systems biology*, 1(1):93–103, 2004.

A. E. C. Ihekwaba, D. S. Broomhead, R. Grimley, N. Benson, M. R. H. White, and D. B. Kell. Synergistic control of oscillations in the NF-kappa B signalling pathway. *Iee Proceedings Systems Biology*, 152(3):153–160, 2005.

P. J. Ingram, M. P. Stumpf, and J. Stark. Network motifs: structure does not determine function. *BMC Genomics*, 7:108, 2006.

C. H. Johnson, M. Egli, and P. L. Stewart. Structural insights into a circadian oscillator. *Science*, 322(5902):697–701, 2008.

I. W. Jolma, O. D. Laerum, C. Lillo, and P. Ruoff. Circadian oscillators in eukaryotes. *Wiley Interdiscip Rev Syst Biol Med*, 2(5):533–49, 2010.

H. Kacser and J. A. Burns. The control of flux. *Symp Soc Exp Biol*, 27:65–104, 1973.

J. D. Kearns, S. Basak, S. L. Werner, C. S. Huang, and A. Hoffmann. IkappaBepsilon provides negative feedback to control NF-kappaB oscillations, signaling dynamics, and inflammatory gene expression. *J Cell Biol*, 173(5):659–64, 2006.

B. N. Kholodenko. Negative feedback and ultrasensitivity can bring about oscillations in the mitogen-activated protein kinase cascades. *Eur J Biochem*, 267(6):1583–8, 2000.

B. N. Kholodenko and M. R. Birtwistle. Four-dimensional dynamics of MAPK information processing systems. *Wiley Interdiscip Rev Syst Biol Med*, 1(1):28–44, 2009.

D. Kimelman and W. Xu. beta-catenin destruction complex: insights and questions from a structural perspective. *Oncogene*, 25(57):7482–7491, 2006.

H. Kitano. Towards a theory of biological robustness. *Mol Syst Biol*, 3:137, 2007.

Y. Kitayama, T. Nishiwaki, K. Terauchi, and T. Kondo. Dual KaiC-based oscillations constitute the circadian system of cyanobacteria. *Genes Dev*, 22(11):1513–21, 2008.

E. Klipp and W. Liebermeister. Mathematical modeling of intracellular signaling pathways. *BMC Neurosci*, 7 Suppl 1:S10, 2006.

R. Krüger and R. Heinrich. Model reduction and analysis of robustness for the Wnt/beta-catenin signal transduction pathway. *Genome Inform*, 15(1):138–48, 2004.

G. Kurosawa and Y. Iwasa. Saturation of enzyme kinetics in circadian clock models. *J Biol Rhythms*, 17(6):568–77, 2002.

Y. K. Kwon and K. H. Cho. Coherent coupling of feedback loops: a design principle of cell signaling networks. *Bioinformatics*, 24(17):1926–32, 2008.

D. H. Le and Y. K. Kwon. A coherent feedforward loop design principle to sustain robustness of biological networks. *Bioinformatics*, 29(5):630–7, 2013.

N. Le Novere, B. Bornstein, A. Broicher, M. Courtot, M. Donizelli, H. Dharuri, L. Li, H. Sauro, M. Schilstra, B. Shapiro, J. L. Snoep, and M. Hucka. Biomodels database: a free, centralized database of curated, published, quantitative kinetic models of biochemical and cellular systems. *Nucleic Acids Research*, 34:D689–D691, 2006.

S. Legewie, N. Blüthgen, and H. Herzel. Quantitative analysis of ultrasensitive responses. *The FEBS journal*, 272(16):4071–9, 2005.

J. C. Leloup. Circadian clocks and phosphorylation: Insights from computational modeling. *Central European Journal of Biology*, 4(3):290–303, 2009.

J. C. Leloup and A. Goldbeter. Modeling the mammalian circadian clock: Sensitivity analysis and multiplicity of oscillatory mechanisms. *Journal of Theoretical Biology*, 230 (4):541–562, 2004.

R. Lev Bar-Or, R. Maya, L. A. Segel, U. Alon, A. J. Levine, and M. Oren. Generation of oscillations by the p53-Mdm2 feedback loop: a theoretical and experimental study. *Proc Natl Acad Sci U S A*, 97(21):11250–5, 2000.

C. Li, M. Donizelli, N. Rodriguez, H. Dharuri, L. Endler, V. Chelliah, L. Li, E. U. He, A. Henry, M. I. Stefan, J. L. Snoep, M. Hucka, N. Le Novere, and C. Laibe. Biomodels database: An enhanced, curated and annotated resource for published quantitative kinetic models. *Bmc Systems Biology*, 4, 2010.

Y. Li and J. Srividhya. Goldbeter-Koshland model for open signaling cascades: a mathematical study. *Journal of mathematical biology*, 61(6):781–803, 2010.

J. C. Locke, A. J. Millar, and M. S. Turner. Modelling genetic networks with noisy and varied experimental data: the circadian clock in Arabidopsis thaliana. *J Theor Biol*, 234(3):383–93, 2005a.

J. C. Locke, P. O. Westermark, A. Kramer, and H. Herzel. Global parameter search reveals design principles of the mammalian circadian clock. *Bmc Systems Biology*, 2:22, 2008.

J. C. W. Locke, M. M. Southern, L. Kozma-Bognar, V. Hibberd, P. E. Brown, M. S. Turner, and A. J. Millar. Extension of a genetic network model by iterative experimentation and mathematical analysis. *Molecular Systems Biology*, 1, 2005b.

L. Ma and P. A. Iglesias. Quantifying robustness of biochemical network models. *BMC Bioinformatics*, 3:38, 2002.

W. Ma, A. Trusina, H. El-Samad, W. A. Lim, and C. Tang. Defining network topologies that can achieve biochemical adaptation. *Cell*, 138(4):760–73, 2009.

N. I. Markevich, J. B. Hoek, and B. N. Kholodenko. Signaling switches and bistability arising from multisite phosphorylation in protein kinase cascades. *J Cell Biol*, 164(3): 353–9, 2004.

O. C. Martin and A. Wagner. Multifunctionality and robustness trade-offs in model genetic circuits. *Biophys J*, 94(8):2927–37, 2008.

B. Mengel, A. Hunziker, L. Pedersen, A. Trusina, M. H. Jensen, and S. Krishna. Modeling oscillatory control in NF-kappaB, p53 and Wnt signaling. *Curr Opin Genet Dev*, 20(6): 656–64, 2010.

L. Michaelis and M. L. Menten. Die Kinetik der Invertinwirkung. *Biochemische Zeitschrift*, 49:333–369, 1913.

J. Monod and F. Jacob. Teleonomic mechanisms in cellular metabolism, growth, and differentiation. *Cold Spring Harb Symp Quant Biol*, 26:389–401, 1961.

S. Nakamura, K. Yokota, and I. Yamazaki. Sustained oscillations in a lactoperoxidase. NADPH and O2 system. *Nature*, 222(5195):794, 1969.

D. E. Nelson, A. E. C. Ihekwaba, M. Elliott, J. R. Johnson, C. A. Gibney, B. E. Foreman, G. Nelson, V. See, C. A. Horton, D. G. Spiller, S. W. Edwards, H. P. McDowell, J. F. Unitt, E. Sullivan, R. Grimley, N. Benson, D. Broomhead, D. B. Kell, and M. R. H. White. Oscillations in NF-kappa B signaling control the dynamics of gene expression. *Science*, 306(5696):704–708, 2004.

L. K. Nguyen. Regulation of oscillation dynamics in biochemical systems with dual negative feedback loops. *Journal of the Royal Society Interface*, 9(73):1998–2010, 2012.

L. K. Nguyen and D. Kulasiri. On the functional diversity of dynamical behaviour in genetic and metabolic feedback systems. *Bmc Systems Biology*, 3, 2009.

B. Novak and J. J. Tyson. Design principles of biochemical oscillators. *Nat Rev Mol Cell Biol*, 9(12):981–91, 2008.

F. Ortega, L. Acerenza, H. V. Westerhoff, F. Mas, and M. Cascante. Product dependence and bifunctionality compromise the ultrasensitivity of signal transduction cascades. *Proc Natl Acad Sci U S A*, 99(3):1170–5, 2002.

P. Paszek, S. Ryan, L. Ashall, K. Sillitoe, C. V. Harper, D. G. Spiller, D. A. Rand, and M. R. White. Population robustness arising from cellular heterogeneity. *Proc Natl Acad Sci U S A*, 107(25):11644–9, 2010.

M. G. Pedersen and A. M. Bersani. Introducing total substrates simplifies theoretical analysis at non-negligible enzyme concentrations: pseudo first-order kinetics and the loss of zero-order ultrasensitivity. *J Math Biol*, 60(2):267–83, 2010.

C. S. Pittendrigh. On temperature independence in the clock system controlling emergence time in Drosophila. *Proc Natl Acad Sci U S A*, 40(10):1018–29, 1954.

C. S. Pittendrigh and P. C. Caldarola. General homeostasis of the frequency of circadian oscillations. *Proc Natl Acad Sci U S A*, 70(9):2697–701, 1973.

A. Plotnikov, E. Zehorai, S. Procaccia, and R. Seger. The MAPK cascades: signaling components, nuclear roles and mechanisms of nuclear translocation. *Biochim Biophys Acta*, 1813(9):1619–33, 2011.

O. Purcell, N. J. Savery, C. S. Grierson, and M. di Bernardo. A comparative analysis of synthetic genetic oscillators. *J R Soc Interface*, 7(52):1503–24, 2010.

J. E. Purvis and G. Lahav. Encoding and decoding cellular information through signaling dynamics. *Cell*, 152(5):945–56, 2013.

Z. Qu and T. M. Vondriska. The effects of cascade length, kinetics and feedback loops on biological signal transduction dynamics in a simplified cascade model. *Physical biology*, 6(1):016007, 2009.

H. Rinne. *Taschenbuch der Statistik*. Wissenschaftlicher Verlag Harri Deutsch GmbH, Frankfurt a.M., 4 edition, 2008.

T. A. Rooney, E. J. Sass, and A. P. Thomas. Characterization of cytosolic calcium oscillations induced by phenylephrine and vasopressin in single fura-2-loaded hepatocytes. *J Biol Chem*, 264(29):17131–41, 1989.

N. F. Ruby, D. E. Burns, and H. C. Heller. Circadian rhythms in the suprachiasmatic nucleus are temperature-compensated and phase-shifted by heat pulses in vitro. *J Neurosci*, 19(19):8630–6, 1999.

P. Ruoff, A. Behzadi, M. Hauglid, M. Vinsjevik, and H. Havas. pH homeostasis of the circadian sporulation rhythm in clock mutants of Neurospora crassa. *Chronobiol Int*, 17(6):733–50, 2000.

P. Ruoff, M. K. Christensen, J. Wolf, and R. Heinrich. Temperature dependency and temperature compensation in a model of yeast glycolytic oscillations. *Biophys Chem*, 106(2):179–92, 2003.

T. Saithong, K. J. Painter, and A. J. Millar. The contributions of interlocking loops and extensive nonlinearity to the properties of circadian clock models. *PLoS One*, 5(11): e13867, 2010a.

T. Saithong, K. J. Painter, and A. J. Millar. Consistent robustness analysis (cra) identifies biologically relevant properties of regulatory network models. *PLoS One*, 5(12):e15589, 2010b.

Kenyi Saito-Diaz, Tony W. Chen, Xiaoxi Wang, Curtis A. Thorne, Heather A. Wallace, Andrea Page-McCaw, and Ethan Lee. The way Wnt works: Components and mechanism. *Growth Factors*, 31(1):1–31, 2012.

C. Salazar and T. Höfer. Competition effects shape the response sensitivity and kinetics of phosphorylation cycles in cell signaling. *Annals of the New York Academy of Sciences*, 1091:517–30, 2006.

C. Salazar and T. Höfer. Multisite protein phosphorylation–from molecular mechanisms to kinetic models. *The FEBS journal*, 276(12):3177–98, 2009.

M. J. Sanderson, P. Delmotte, Y. Bai, and J. F. Perez-Zogbhi. Regulation of airway smooth muscle cell contractility by ca2+ signaling and sensitivity. *Proc Am Thorac Soc*, 5(1):23–31, 2008.

J. Schaber, R. Baltanas, A. Bush, E. Klipp, and A. Colman-Lerner. Modelling reveals novel roles of two parallel signalling pathways and homeostatic feedbacks in yeast. *Molecular systems biology*, 8:622, 2012.

C. G. Schipke, A. Heidemann, A. Skupin, O. Peters, M. Falcke, and H. Kettenmann. Temperature and nitric oxide control spontaneous calcium transients in astrocytes. *Cell Calcium*, 43(3):285–95, 2008.

H. Shankaran, D. L. Ippolito, W. B. Chrisler, H. Resat, N. Bollinger, L. K. Opresko, and H. S. Wiley. Rapid and sustained nuclear-cytoplasmic ERK oscillations induced by epidermal growth factor. *Mol Syst Biol*, 5:332, 2009.

G. Shinar and M. Feinberg. Structural sources of robustness in biochemical reaction networks. *Science*, 327(5971):1389–91, 2010.

G. Shinar, J. D. Rabinowitz, and U. Alon. Robustness in glyoxylate bypass regulation. *PLoS Comput Biol*, 5(3):e1000297, 2009.

K. Sneppen, M. A. Micheelsen, and I. B. Dodd. Ultrasensitive gene regulation by positive feedback loops in nucleosome modification. *Molecular systems biology*, 4:182, 2008.

J. Sneyd and J. F. Dufour. A dynamic model of the type-2 inositol trisphosphate receptor. *Proc Natl Acad Sci U S A*, 99(4):2398–403, 2002.

J. Sneyd and M. Falcke. Models of the inositol trisphosphate receptor. *Prog Biophys Mol Biol*, 89(3):207–45, 2005.

J. Sneyd, K. Tsaneva-Atanasova, D. I. Yule, J. L. Thompson, and T. J. Shuttleworth. Control of calcium oscillations by membrane fluxes. *Proc Natl Acad Sci U S A*, 101(5): 1392–6, 2004.

A. Sorkin and M. von Zastrow. Signal transduction and endocytosis: Close encounters of many kinds. *Nature Reviews Molecular Cell Biology*, 3(8):600–614, 2002.

J. Stelling, E. D. Gilles, and 3rd Doyle, F. J. Robustness properties of circadian clock architectures. *Proc Natl Acad Sci U S A*, 101(36):13210–5, 2004a.

J. Stelling, U. Sauer, Z. Szallasi, 3rd Doyle, F. J., and J. Doyle. Robustness of cellular functions. *Cell*, 118(6):675–85, 2004b.

R. Steuer. Computational approaches to the topology, stability and dynamics of metabolic networks. *Phytochemistry*, 68(16-18):2139–51, 2007.

R. Steuer, S. Waldherr, V. Sourjik, and M. Kollmann. Robust signal processing in living cells. *Plos Computational Biology*, 7(11):e1002218, 2011.

I. Stoleriu, F. A. Davidson, and J. L. Liu. Quasi-steady state assumptions for non-isolated enzyme-catalysed reactions. *Journal of mathematical biology*, 48(1):82–104, 2004.

R. Straube. Sensitivity and robustness in covalent modification cycles with a bifunctional converter enzyme. *Biophysical journal*, 105(8):1925–33, 2013.

J. Stricker, S. Cookson, M. R. Bennett, W. H. Mather, L. S. Tsimring, and J. Hasty. A fast, robust and tunable synthetic gene oscillator. *Nature*, 456(7221):516–9, 2008.

S. C. Sun. Non-canonical NF-kappaB signaling pathway. *Cell research*, 21(1):71–85, 2011.

M. H. Sung, L. Salvatore, R. De Lorenzi, A. Indrawan, M. Pasparakis, G. L. Hager, M. E. Bianchi, and A. Agresti. Sustained oscillations of NF-kappaB produce distinct genome scanning and gene expression profiles. *PLoS One*, 4(9):e7163, 2009.

S. E. Szedlacsek, M. L. Cardenas, and A. Cornish-Bowden. Response coefficients of interconvertible enzyme cascades towards effectors that act on one or both modifier enzymes. *European journal of biochemistry / FEBS*, 204(2):807–13, 1992.

S. W. Teng, S. Mukherji, J. R. Moffitt, S. de Buyl, and E. K. O'Shea. Robust circadian oscillations in growing cyanobacteria require transcriptional feedback. *Science*, 340 (6133):737–40, 2013.

R. Thomas and M. Kaufman. Conceptual tools for the integration of data. *Comptes rendus biologies*, 325(4):505–14, 2002.

W.R. Thompson. On confidence ranges for the median and other expectation distributions for populations of unknown distribution form. *Ann. Math. Statist.*, 7:122–128, 1936.

C. D. Thron. Mathematical analysis of binary activation of a cell cycle kinase which down-regulates its own inhibitor. *Biophys Chem*, 79(2):95–106, 1999.

X. J. Tian, X. P. Zhang, F. Liu, and W. Wang. Interlinking positive and negative feedback loops creates a tunable motif in gene regulatory networks. *Physical review. E, Statistical, nonlinear, and soft matter physics*, 80(1 Pt 1):011926, 2009.

M. Tigges, T. T. Marquez-Lago, J. Stelling, and M. Fussenegger. A tunable synthetic mammalian oscillator. *Nature*, 457(7227):309–12, 2009.

J. E. Toettcher, C. Mock, E. Batchelor, A. Loewer, and G. Lahav. A synthetic-natural hybrid oscillator in human cells. *Proc Natl Acad Sci U S A*, 107(39):17047–52, 2010.

T. Y. C. Tsai, Y. S. Choi, W. Z. Ma, J. R. Pomerening, C. Tang, and J. E. Ferrell. Robust, tunable biological oscillations from interlinked positive and negative feedback loops. *Science*, 321(5885):126–129, 2008.

A. R. Tzafriri and E. R. Edelman. Quasi-steady-state kinetics at enzyme and substrate concentrations in excess of the michaelis-menten constant. *Journal of Theoretical Biology*, 245(4):737–748, 2007.

S. B. Van Albada and P. R. Ten Wolde. Enzyme localization can drastically affect signal amplification in signal transduction pathways. *Plos Computational Biology*, 3(10):1925–1934, 2007.

A. Wagner. Circuit topology and the evolution of robustness in two-gene circadian oscillators. *Proc Natl Acad Sci U S A*, 102(33):11775–80, 2005.

A. K. Wilkins, P. I. Barton, and B. Tidor. The Per2 negative feedback loop sets the period in the mammalian circadian clock mechanism. *Plos Computational Biology*, 3 (12):2476–2486, 2007.

A. K. Wilkins, B. Tidor, J. White, and P. I. Barton. Sensitivity analysis for oscillating dynamical systems. *Siam Journal on Scientific Computing*, 31(4):2706–2732, 2009.

J. Wolf, S. Becker-Weimann, and R. Heinrich. Analysing the robustness of cellular rhythms. *Syst Biol (Stevenage)*, 2(1):35–41, 2005.

J. V. Wong, B. Li, and L. You. Tension and robustness in multitasking cellular networks. *PLoS Comput Biol*, 8(4):e1002491, 2012.

W. W. Wong, T. Y. Tsai, and J. C. Liao. Single-cell zeroth-order protein degradation enhances the robustness of synthetic oscillator. *Mol Syst Biol*, 3:130, 2007.

Y. Wu, X. Zhang, J. Yu, and Q. Ouyang. Identification of a topological characteristic responsible for the biological robustness of regulatory networks. *PLoS Comput Biol*, 5 (7):e1000442, 2009.

L. Xu and Z. Qu. Roles of protein ubiquitination and degradation kinetics in biological oscillations. *PLoS One*, 7(4):e34616, 2012.

I. Yamazaki, K. Yokota, and R. Nakajima. Oscillatory oxidations of reduced pyridine nucleotide by peroxidase. *Biochemical and Biophysical Research Communications*, 21 (6):582–6, 1965.

E. Zamora-Sillero, M. Hafner, A. Ibig, J. Stelling, and A. Wagner. Efficient characterization of high-dimensional parameter spaces for systems biology. *Bmc Systems Biology*, 5:142, 2011.

Q. Zhang, S. Bhattacharya, and M. E. Andersen. Ultrasensitive response motifs: basic amplifiers in molecular signalling networks. *Open biology*, 3(4):130031, 2013.

Acknowledgements

This work would not have been possible - or much more difficult to bring all together - without people who also decisively enriched my time as PhD student.

First of all, there are to mention all present and former members of the lab of Jana Wolf at the Max-Delbrück-Center of Molecular Medicine. We had many fruitful discussions, and I took lots of ideas and encouragement from talking to you. In particular, Jana Wolf gave me critical advice and supported me in every situation, and I would like to thank her for all of it. In addition, I would like to thank Bente Kofahl who carefully read my manuscript and thus helped me formulating my thoughts more precisely and understandably, and who last but not least always lent me an ear whenever necessary. Thanks to Dorothea Busse who read parts of the manuscript and helped improving it. Antonio Politi supported me advancing my project during his time at the Wolf lab.

Thanks to Edda Klipp as my supervisor from the Humboldt-University. She invited me for talks in her group and hence enabled stimulating discussions with parts of the theoretical biology community from the Humboldt-University and Charité - and she rendered the bureaucracy in the context of promotion as unbureaucratic as possible. Furthermore, I would like to thank Ralf Steuer for constructive advice concerning especially the employed methods, and for inspiring discussions.

Despite not playing a role for the particular content of this thesis, I would like to mention also Vanessa Schmidt from the Willnow lab and Buket Yilmaz from the Scheidereit lab, both at the MDC, who introduced me to their interesting branches of biological research and called my attention to the complexity and needs of more wet-lab-based approaches.

Weiterhin spielen auch andere Menschen bei der Erstellung dieser Arbeit eine Rolle - wenige direkt, einige indirekt, aber dafür umso mehr. Zunächst möchte ich Frauke Söhler danken, die dem Manuskript im Ganzen zu deutlich weniger Fehlern verholfen hat. Weiterhin gilt großer Dank meinen Eltern - ihr habt mir nicht nur in dieser Zeit, sondern bereits in meinem ganzen Leben immer Beistand geleistet, und seid auch spontan eingesprungen, wenn es nötig war. Schön, dass es Euch gibt! Ganz zum Schluss möchte ich außerdem Anco Baum danken. Dadurch, dass er mich immer vorbehaltlos unterstützt hat und für mich da war (zum Glück auch nicht nur, wenn nötig), hat er eine sehr große Rolle für das Gelingen dieser Dissertation gespielt - Worte können kaum ausdrücken, wieviel Dank ihm dafür gebührt.